THE NATURE-STUDY IDEA

A VOLUME IN THE SERIES

THE LIBERTY HYDE BAILEY LIBRARY

Edited by John Linstrom

A list of titles in this series is available at cornellpress.cornell.edu.
Additional resources are available at the companion website
lhbaileyproject.com.

Liberty Hyde Bailey, right, examines a sandwich with Anna Botsford
Comstock, opposite, at the first Tompkins County School Picnic, May 26, 1905. They
appear to be conversing with county schoolteachers. Courtesy of the Division of
Rare and Manuscript Collections, Cornell University Library.

The Nature-Study Idea

And Related Writings

Liberty Hyde Bailey
Edited by John Linstrom
With a Foreword by David W. Orr

Comstock Publishing Associates
an imprint of
Cornell University Press
Ithaca and London

Thanks to generous funding from the American Horticultural Society, the Antioch University Nature-Based Early Childhood Program, the Brandwein Institute, Ann Habicht, #NatureForAll/the International Union for Conservation of Nature Commission on Education and Communication, and Jack Padalino, the ebook editions of this book are available as open access volumes through the Cornell Open initiative.

AMERICAN
HORTICULTURAL
SOCIETY

BRANDWEIN
I N S T I T U T E

First edition published 1903. Fourth edition 1911.
Edition with related writings first published 2023 by Cornell University Press

Library of Congress Cataloging-in-Publication Data

Names: Bailey, L. H. (Liberty Hyde), 1858–1954, author. | Linstrom, John, 1987– editor. | Orr, David W., 1944– writer of foreword. | Williams, Dilafruz R., 1949– writer of added commentary. | Bailey, L. H. (Liberty Hyde), 1858–1954. Nature-study idea Fourth edition, revised.
Title: The nature-study idea : and related writings / Liberty Hyde Bailey ; edited by John Linstrom ; with a foreword by David W. Orr.
Description: Ithaca : Comstock Publishing Associates, an imprint of Cornell University Press, 2023. | Series: The Liberty Hyde Bailey library | Includes the text of the "final, 1911 fourth edition of The Nature-Study Idea in its entirety, respecting all the changes Bailey made over the course of four editions to one of his most important books"— page xii of ECIP galley. | Includes bibliographical references and index.
Identifiers: LCCN 2023024138 (print) | LCCN 2023024139 (ebook) | ISBN 9781501772610 (hardcover) | ISBN 9781501773952 (paperback) | ISBN 9781501772627 (pdf) | ISBN 9781501772634 (epub)
Subjects: LCSH: Bailey, L. H. (Liberty Hyde), 1858–1954. Nature-study idea. | Nature study.
Classification: LCC QH51 .B16 2023 (print) | LCC QH51 (ebook) | DDC 508—dc23/eng/20230722
LC record available at https://lccn.loc.gov/2023024138
LC ebook record available at https://lccn.loc.gov/2023024139

This edition is dedicated to the memory of Jane L. Taylor,
champion of children's gardens, of outdoor learning,
and of Bailey's work, including his message
that teachers not be afraid to teach;

to my mother,
Rebecca Johnson Linstrom,
teacher and nurturer of children's whole selves
and embodiment of "the artistic expression of life";

and to Chloe,
who enters the world with the coming spring
and for whom, as the years ripen, I hope I earn the name
of teacher, father, and friend.

The power that moves the world is the power of the teacher.
Liberty Hyde Bailey, *The Nature-Study Idea*

CHILD'S REALM

A little child sat on the sloping strand
    Gazing at the flow and the free,
Thrusting its feet into the golden sand,
    Playing with the waves and the sea.

        I snatched a weed that was tossed on the flood
            And unravelled its tangled skeins;
        And I traced the course of its fertile blood
            That lay deep in its meshèd veins;

        I told how the stars are garnered in space
            How the moon on its course is rolled;
        How the earth is hung in its ceaseless place
            As it whirls in its orbit old.

The little child paused with its busy hands
    And gazed for a moment at me,
Then it dropped again to its golden sands
    And played with the waves and the sea.

Liberty Hyde Bailey, from *Wind and Weather*, 1916, p. 119

CONTENTS

Illustrations

FOREWORD

In *The Nature-Study Idea* (1903), Liberty Hyde Bailey proposed adding the study of nature to school curricula not just to instill knowledge about the natural world but as a method to awaken the child's spirit and inform their worldview.[1] The aim was to enable the children to develop a thoughtful and competent love for nature that grew from their curiosity about the natural world. "The first essential," he wrote, "is an intense love of nature," and all else would follow in due course, including scientific knowledge and ethical awareness (Part I, Chapter I, this volume; hereafter, "I.I"). That intense love of nature grew best out of doors and on the child's terms. Teachers facilitated, a child's curiosity prevailed over curriculum, and disciplines were boundaries to be crossed. The point was to nurture a wider context for living in harmony with nature. Nature-study, in Anna Botsford Comstock's words, was to "be so much a part of the child's thought and interest that it will naturally form a thought core for other subjects quite unconsciously on his [or her] part."[2] The nature-study movement aimed to enable an individual to use their senses, keep their eyes open, and awaken to "the beauty as well as to the wonders which are there."[3]

Bailey's view of nature was a precursor to what was later called "deep ecology," in which nature was given profound consideration, if not legal rights. The "notion that all things were made for man's special pleasure," he wrote, "is colossal self-assurance" (II.III). As an antidote to the triumphalism of his time, Bailey proposed "that all people, or as many of them as possible, shall have contact with the earth and that the earth righteousness shall be abundantly taught."[4] That curriculum, however, began in humility tempered by reverence for life—a precursor as well to Albert Schweitzer's philosophy and that of the tribe of environmental ethicists to come.

The nature-study movement aimed beyond children to farmers and the improvement of rural life. "No thoroughly good farming," he wrote, "is possible without this same knowledge and outlook. Good farmers are good naturalists" (I.VII). Further, he regarded "extend[ing] the agricultural applications of nature-study" as the "special mission" of an agricultural college, including his home institution, Cornell University (I.VII). Bailey's views about farming and rural improvement also preceded those of others, including Louis Bromfield and Aldo Leopold.

The nature-study movement that "began to take form [. . .] from 1884 to 1890" was roughly coincidental in time with the early work of the philosopher, John Dewey (I.II). Bailey and Dewey both intended to reorient education to place and locality, and both considered the home and neighborhood as the foundation of democracy.[5] At the time, most Americans lived on farms or in small towns and knew the rigors of rural life. America, however, was on the move, busy manifesting its destiny, "winning the West," conquering the few remaining recalcitrants, building cities, taking the first steps to an overseas empire, industrializing with a vengeance, building vast corporations, and, for a few, amassing huge fortunes. Those like John Muir and John Burroughs, and the rare politicians like Teddy Roosevelt and Gifford Pinchot who had second thoughts about the juggernaut sweeping across the American land, put up little more than speedbumps on the road to our current predicament. The Country Life Commission, appointed by Roosevelt in 1908 and chaired by Bailey, sought to focus on the needs of rural people and towns left behind. Despite several notable successes, including the Smith-Lever Act of 1914, which created a national agricultural extension program through the land-grant universities, the country life movement was no match for capital-intensive, large-scale agriculture. Capitalism swept across rural America like a plague, devouring people, land, forests, wildlife, waters, small farms, and once-vibrant small towns alike. Urban sprawl, commercialism, and interstate highways did the rest. As a result, the rural life movement foundered, in Paul Sears's words, on "our utter failure to see the connection between the word 'Conservative' and the word 'Conservation.' "[6]

The idea of nature education, however, has not perished; it was too good and too necessary to go away. Bailey, Comstock, and others are largely forgotten, but their legacy and ideas live in the work of environmental educators, environmental organizations, and local land preservation organizations and in organizations such as the Children and Nature Network, inspired by Richard Louv.[7] But the challenge is greater than Bailey and his contemporaries could have foreseen. Most children now grow up in urban areas and live indoors, all too often addicted to their smartphones, their free time filled with the internet and a thousand distractions that capture their attention and minds.[8] As a result, fewer now spend much time out of doors or live in places where contact with nature is routine, necessary, and instructive.

When Bailey passed from this Holy Earth in 1954 at the age of 96, CO_2 in the atmosphere was 313 ppm. As I write (March 2022), it is near 420 ppm. To that number we should add another 50–75 ppm of other heat-trapping gases measured in CO_2 equivalent units, putting us close to 500 ppm higher than any time in the past two to five million years. In 1954, the United States stood astride the world like that proverbial colossus. Now, not so much. Bailey's early years were the era of the "robber barons." We too have robber barons, just as predatory or even worse. A half dozen U.S. oligarchs have more wealth than the bottom 50% of the population, and the gap between the richest and the rest continues to widen. In 1954 there were ~5.6 million farms in the United States, and the average farm size slightly more than two hundred acres. Presently we have two million farms, but production is dominated by eighty thousand mega farms. The diversified small farm of Bailey's day cannot compete with highly subsidized, chemicalized, and capital-intensive agribusiness. Perhaps one-third or more of the topsoil on American farmland in 1900 has since washed or blown away, but no one knows for certain. Excess fertilizer runoff down the Mississippi has created a dead zone in the Gulf of Mexico the size of New Jersey. Another dead zone is growing between our ears, a national deficit of the kind of knowledge and sensibility that was the core of nature-study. And not the least, our oil-soaked democracy seems to have stalled out, our institutions corrupted by too much unaccounted money and elected officials with too much ambition and too little integrity. Too much venom, too little kindness. We are vexed and paralyzed, unable to solve even the most basic public problems.

Bailey, a man who thought a great deal about connectedness, would have noticed those connections between land health, human well-being, oil, rapid climate change, and the shabbiness of our public affairs. In *What is Democracy?* he wrote, "A nation of selfish individuals is never a democracy. A democratic society is impossible until its population is possessed of the spirit of helpfulness to others."[9] Like Thomas Jefferson, Bailey believed that "democracy rests on the land" but never on a landed aristocracy, let alone a land-owning corporatocracy.[10] Like Tocqueville, Bailey regarded the "habits of heart"—the inclinations and dispositions nurtured on farms—as one of "the great sources of citizenship" and the foundation for "a permanent society." "The farmer," he wrote, "is the fundamental fact in a democracy."[11] But he regarded growing food for urban society as secondary to the role of farms and rural communities in providing a steady flow of young people recruited into our national life—a citizenry disciplined by hard work and frugality; knowledgeable about soils, animals, wildlife, weather, and water; and adept at solving practical problems with home-grown ingenuity.[12] Bailey's vision also included partnerships between the city, country towns, and farms that provided a steady supply of food while strengthening common bonds.[13]

Troubled by the destruction wrought by World War I, Bailey proposed an alternative to worldwide militarism in the form of a "cooperative effort for the public good, rather than for the public destruction [. . . and] the shameful wounding of the planet."[14] He proposed instead the "constructive occupancy" and "reconstruction of the earth" with rural people in the lead and organized as "a Society of the Holy Earth. Chapters and branches it may have, but its purpose is not to be [an] organization [. . .]. Its principle of union will be the love of the Earth, treasured in the hearts of men and women."[15] Later, those vague ideas became manifest in the Peace Corps, VISTA, and other organizations providing opportunities for public service for the young and idealistic.

What can be made of this remarkable man as a scientist, writer, and advocate? More important, what should we do with his legacy in our own time? First, Bailey is among the most prescient critics of industrial society, including Henry David Thoreau, George Perkins Marsh, and especially John Wesley Powell, who warned that settlement of arid regions of the West should be organized by watersheds rather than the hydrological fantasies of land speculators. Second, Bailey cannot rightly be dismissed as a quaint anachronism, a throwback to some long-gone era. I think it is more accurate to describe Bailey as a "throw forward," a visionary who saw more clearly than most what must be done to make America more than a trial balloon, as Aldo Leopold put it. Bailey understood the importance of farming and farm life, and the necessity of vibrant and prosperous rural areas, for the durability of any civilization. He knew how important kindness, neighborliness, and cooperation were to the democratic temperament. He knew that democratic societies can be sustained only by ecologically literate, practically competent, and community-minded people. Nature-study, in other words, was not just a curriculum for children but an essential part of democracy, revitalized rural areas, and a prosperous and permanent agri-culture.[16]

I believe that those traits have increased in value, not the least because our generation and those to come face a rapidly warming climate and the prospect of cascading systemic failures that threaten the ecological and material underpinnings of civilization. The man revealed in his life and writings would not have equivocated in the face of the long climate emergency ahead. Rather, I think he would have set about to rebuild the frayed foundations of rural America and worked to make sustainable rural prosperity a reality. It is not difficult to imagine a Liberty Hyde Bailey now enlisting liberal arts colleges and land-grant universities, including his own, to the cause of "universal service" extended to include future generations, and in that effort giving birth to a post-extractive agriculture that would not mine soils, groundwater, people, or communities; an agri-culture that would sequester carbon in soils and would dependably render sunlight into plant tissue, animal flesh, and electrons; an agri-culture built not on corporate power and high finance, but ingenuity, devotion, skill, and the authentic patriotism of people who know that the Earth is indeed Holy and that it can be redeemed only by loving care and practical competence.

David W. Orr

INTRODUCTION

> A book like [*The Principles of Agriculture*] should be used only
> by persons who know how to observe. The starting-point in the
> teaching of agriculture is nature-study,—the training of the power to
> actually see things and then to draw proper conclusions from them.
> Into this primary field the author hopes to enter; but the present need
> seems to be for a book of principles designed to aid those who know
> how to use their eyes.
>
> Liberty Hyde Bailey, preface to *The Principles
> of Agriculture* (1898), p. viii

Liberty Hyde Bailey would realize his ambition to enter "this primary
field" just five years after writing that preface to *The Principles of Agriculture* in what would prove to be one of his most influential and enduring texts, *The Nature-Study Idea*: a book not only for "those who
know how to use their eyes," but one that would help others to see, as
well. A rising leader and visionary in the field of agricultural education—
a new concept then emerging from the series of land-grant college acts
meant to make higher education relevant and accessible to the majority of
Americans who were still farmers at that time—Bailey believed that a better, more sustainable future for agriculture and rural life would not come
about simply through scientific investigation and improved farming methods. These methods would be needed, such as crop rotation to sustain
and build soil fertility, and he outlined them in books such as *The Principles of Agriculture* (it comes as no surprise, in fact, that a full third of that
book was devoted to the health of the soil). But more fundamental and
more pressing than any program of scientific advancement would be an

educational model oriented toward the complexity of the natural world and rooted in a sense of wonder, an outlook best nurtured in the early elementary years but important for all of us, throughout our lives. For this reason, nature-study was much more than science, even as it embraced a scientific outlook; in fact, Bailey argued, nature-study "is not science. It is not knowledge. It is not facts. It is spirit. It is an attitude of mind. It concerns itself with the child's outlook on the world" (Part I, Chapter I; hereafter, I.I). He believed that such an "outlook to nature," as he would call it in the title of his next major book, would become the best safeguard of a more resilient, sustainable agriculture.

Today we understand that agricultural crises, like the horrifically rapid loss of our planet's topsoil (that precious living realm of microbial ecosystems that remains so mysterious and so crucial to our ability to feed ourselves as a species) are inextricably bound to the even larger crisis of climate change.[1] And as David W. Orr notes in his foreword to this volume, our ability to collectively address these large, systemic crises relates directly to the health of democratic systems around the world—systems that are also undergoing unprecedented strain, sometimes for good reasons but too often due to the oligarchic ambitions of the few. How do we strengthen democracy (a state of society that Bailey felt had never been fully reached but that we should constantly strive for)[2] and at the same time mobilize every possible resource toward the remediation of climate change and a truly regenerative future? Scientific facts alone will not push the massive political shifts and reallocation of resources that we need. What progress we have seen in recent years has come about when those facts were mobilized by an increasingly powerful cultural movement demanding change—a diverse and coalitional movement decades in the making that needs to accelerate right now.

For Bailey, the emergent science of evolution provided a striking affirmation of "the simple wisdom of the fields" that he felt he had grown up with in nineteenth-century rural Michigan: the awareness that, as he writes in *The Nature-Study Idea*, "all things are of kin" (II.I), and that this awareness of our deep familial kinship with all life requires of us a new kind of interspecies ethics, a responsibility to the more-than-human world. In *The Holy Earth* (1915), he would memorably describe the need for "a new hold" on our place in nature, on a planet that "is not exclusively man-centred; it is bio-centric." This affirmation from the lessons of evolution still speaks to us today, and the science of climate change only intensifies our awareness of our deep contingency in the web of life (that the earth "is not exclusively man-centred") at the same time that it drives home just what a different order of magnitude the impact of our species on the planet is than that of any other. For Bailey, this relational awareness had the potential, once society was truly awakened to it, of revolutionizing every aspect of our social life, of reintegrating country and city, of fitting our homes, farms, and communities to a finer-tuned sympathy with nature. A leader and an organizer himself, particularly from his seat as Experiment Station Director and later the founding Dean of the College of Agriculture at Cornell University, he then set about to think through how this awareness could actually be spread throughout a culture, in a process identified by Bailey scholars Paul A. Morgan and Scott J. Peters as "worldview transition."[3] Bailey believed that the work of such worldview transition would fundamentally rest upon a renewed education—an education rooted not in books, or in merely learning the names of things or collections of facts, but in "the simple wisdom of the fields," experienced directly and with the whole child in mind. Nature-study wasn't just natural science adapted for young children; it was part of "a soul-movement" (II.I).

Such a soul-movement would strengthen the bonds of democracy, both through bringing people together (with each other and with our nonhuman kin) and through challenging the injustices that have kept us apart. A democratizing education would mobilize knowledge in ways that empower communities to adapt and, when necessary, to develop a public mandate for change. Bailey knew that we would need more than knowledge or facts to reform our outlook to nature, and he also knew that the natural world itself had much more to teach us than scientific knowledge alone. Science is a human institution, after all; the more-than-human world speaks a more capacious set of languages, and nature-study would open the senses to these multiple ways of coming to know "the simple wisdom of the fields," informed but not limited by the formal sciences. The humanities and the arts would point the way to "the poetic interpretation of nature" (II.VI) and the development of an ethic of environmental stewardship, from the individual to the social level. No academic discipline should be elevated above the others in the education of the whole child—no snappy acronyms or corporate educational marketing campaigns should restrict the child's exploration (he writes of the danger of "catch-words" in The Science Element in Education, in Related Writings, this volume)—and all of these disciplinary endeavors would come to nothing, pedagogically, if they failed to root themselves in firsthand, experiential learning in the real world, out of doors. Nature would provide the check to human (and disciplinary) arrogance.

Moreover, Bailey argued, "The outlook to nature is the outlook to optimism."[4] Nature-study would be the means to raise happier, as well as more well-rounded and more ecologically minded, children. Teachers would find it uplifting as well, and *The Nature-Study Idea* owed its success in its own time largely to its lasting inspirational and instructive value to the art of teaching.

The book staked out the intellectual territory for an influential and pedagogically powerful vision for educational reform, and that vision continues to challenge current trends. Bailey combined an emphasis on experiential learning with a distinctly ecological understanding that the best way to learn about the world, how it works, and how we fit within it, is by observing plants and animals in their own habitats. After all, as he points out in The Humanistic Element in Education (in Related Writings, this volume), "Man is as much a part of nature as is a pigeon or a trillium." He had known since his childhood on the farm that children were active agents in their ecosystems—and so were teachers, and so were the gardens that began rapidly to proliferate across turn-of-the-century schoolyards under the influence of the nature-study movement. Bailey was never one to impose an artificial divide between the human world and the rest of nature, instead always insisting on the human place as a species embedded within natural systems, with human actions not only influencing but also influenced by surrounding environments (for better or worse, domesticated or wild). The child should feel and be encouraged to explore that sense of relationship with and embeddedness in her more-than-human neighborhood, he insisted, and that neighborhood should extend to untamed nature (how many birds, insects, and weeds does even the city child pass each day on the way to school!) as well as to the working landscapes of the community. In Bailey's primary sphere of concern, and that of most Americans a century ago, those working landscapes were the farms and gardens of the open country, but he also argued that nature-study should not be confined to rural areas only.

The goal of this ecological (Bailey would simply say "natural") education was not primarily to produce academic specialists in any field but rather "the establishment of a living sympathy with everything that is" (What Is Nature-Study?, in Related Writings, this volume) in order "to enable every person to live a richer life, whatever his business or profession may be" (I.I). Such a goal stands in stark contrast to persisting obsessions with the merely testable, quantifiable, and measurable. In his 1909 revision of the book, Bailey responded to requests for "statistics" quantifying the impact of nature-study, writing that the assumption that such numbers could even be given

> misses the very purpose of the nature-study movement, which is to set pupils at work informally and personally with the objects, the affairs and phenomena with which they are in daily contact. There are very many teachers and very many schools, and very many pupils, who have a new outlook on life as the result of nature-study work; but if I could give a statistical measure of the nature-study movement, I should consider the work to have been a failure, however large the figures might be. (I.I)

If he were writing today, Bailey would undoubtedly have something to say about the pedagogical impacts of high-stakes testing and curriculum cramming, motivated by a cultural obsession with "statistical measures" that so often leave little room for the development of curiosity and discovery— in other words, for education, for the development of the child, a result that cannot be captured numerically.

In beginning with the common things of the neighborhood, which are relevant to the child's life because they constitute the child's world, nature-study works against superficiality in education. By basing itself in the concrete and exploratory experiences of the child, Bailey argued that it also provided a more reliable model for understanding the world than the second-hand information in textbooks, which at the time were marketed aggressively to teachers, often by dubious salespeople interested in turning a quick profit on the burgeoning field of public schooling, in much the same way that web-based educational applications are often marketed to teachers and administrators today.[5] "Nature, not books," was the famous dictum of nature-study teachers taken from the work of Louis Agassiz, and nature was free to all right outside the schoolhouse door. Simple, entertaining pamphlets to help direct the activities and point of view were free to whomever requested them from Cornell, but they always sought to send the teacher and students away from the pamphlets themselves and out into the world. And anyway, the technological tools that today can connect children to seemingly infinite information (and disinformation), like the books of Bailey's time, will always run the risk of alienating students, rather than nurturing productive connections, if they are not accompanied by a more foundational grounding in the world the children actually inhabit.

The potential ecological impact of raising each generation to be more sensitive than their parents to the realities of the natural world has grown only more significant for us on a climate-changed planet. In 1903, Bailey already saw such work to be foundational to a culturally and ecologically sustainable future. In response to a historical moment in which farmers were being pushed off the land by a series of droughts and a set of economic policies that prioritized consumers over producers—problems more acute today than then and less well appreciated—Bailey was able to marshal state funding for a nature-study program, of all things, in the belief that the mere dissemination of information alone would not solve the underlying problems facing agriculture. The future would bring more droughts, and in the meantime economic policy would need to be rewritten, but by cultivating a sense of curiosity and wonder and by empowering students (many of whom, in that day, *would* go on to become farmers) to investigate problems for themselves, nature-study provided the necessary foundation for a healthy "rural civilization," as he sometimes called it, in which better methods would be discovered every day on working farms and in which farm life would be not only economically viable but intellectually stimulating, and thereby fulfilling. Nature-study planted the seeds for improved farming, but more importantly, it also planted the seeds for a more rewarding and fuller life on the land.

That this fuller life was meant for everyone was implied by the grassroots character of the nature-study movement itself. Because they emerged primarily from the work of teachers in the public schools, nature-study methods were developed to be simple, affordable, time effective, and accessible to the greatest number of students. Bailey takes pains in *The Nature-Study Idea* to trace the movement's evolution and cite the work of many individuals working in their local communities for the good of their children—ordinary educators and administrators, many of whom would otherwise have been lost to history. Moreover, it was a movement led largely by women, and Bailey appointed women, including most notably Anna Botsford Comstock, to lead the effort at Cornell in prominent positions at a time when university professors were almost exclusively men. In the South, nature-study was taken up by Black educators like Booker T. Washington and George Washington Carver, who argued for it largely along the same lines that their white northern counterparts did: as a pedagogical approach that encouraged independent thinking, exploration, and sympathy with the natural world, leading children to greater self-sufficiency and intellectual satisfaction with rural life. And nature-study quickly became popular in cities like Chicago and New York, where vacant lots were transformed into massive school gardens and "field trips" (an invention of the nature-study movement) provided engaging opportunities for learning as well as escapes from the stiflingly overcrowded classrooms characteristic of urban public schools of the era. As a democratizing ideal, nature-study was understood to benefit anyone lucky enough to encounter it in school or at home, and it was the mission of the movement's leaders that nature-study reach as many children as possible. (More on this history and how Bailey fits into it in my essay "It Is Spirit," this volume.)

Bailey shared this belief that nature-study was universally needed in his day, both to provide children with a lifelong balm from the increasing complications of modern life and to work as a corrective to the domineering mind-set that placed humans apart from and above nature—a mind-set that led to what he would later call the "habit of destruction."[6] He could not have foreseen the extent of that destruction in those early days of climate science, although the realities of species extinction were increasingly clear to him. The last passenger pigeon would die eleven years after the publication of *The Nature-Study Idea*, and he could still remember when their migrations would darken the skies of his childhood farm home for days. Moreover, Bailey could see, more clearly than many today, that questions of ecology and conservation were inextricable from questions of education, community organization, the relationship of country to city, and the functioning of democracy. Bailey would write more about each of these topics over the course of his career as a leading agrarian thinker and reformer, but he always maintained that the root of all possible reform in any of these arenas would be nature-study.

This edition of *The Nature-Study Idea* seeks to reintroduce Bailey's classic work to best speak to the needs of contemporary readers—teachers, parents, students of education and of the environment, activists, and scholars of the many intersecting fields that Bailey's work engages. David W. Orr's foreword speaks to the urgency of nature-study in the present moment of climate catastrophe and the fraying of democracy, and Dilafruz R. Williams's essay puts Bailey's words into the context of current pedagogical discourses in the field of education. Following these opening essays is a lengthier historical sketch, titled " 'It Is Spirit': The Genesis of *The Nature-Study Idea*," which tells the story of Bailey's educational philosophy and many of the influences that went into his book, from a beloved childhood teacher to a college mentor with whom he would come into sharp disagreement. Then, following a note on the texts, we present the final, 1911 fourth edition of *The Nature-Study Idea* in its entirety, respecting all the changes Bailey made over the course of four editions to one of his most important books. The text features extensive endnotes for those interested in engaging Bailey's words more deeply, from short biographical sketches of individuals named (who range from classical Greek philosophers to rural school principals) to minor passages eliminated from the first edition in Bailey's later revisions. After the full text of the fourth edition comes a series of larger sections cut from the first edition, which, along with the shorter passages supplied in the endnotes, makes this text the most complete edition of the work available. Then a section of contemporary book reviews of *The Nature-Study Idea* helps fill in the picture of the breadth and quality of the book's reception, both in its first edition with Doubleday and the revised third edition with Macmillan, representing perspectives ranging from those of literary critics to education professionals and sociologists. Finally, a selection of Related Writings provides a number of important, shorter nature-study texts written by Bailey, spanning the years 1896 to 1918, that help to fill in the picture of Bailey's nature-study idea and its evolution over time, from the founding of Cornell's nature-study work, through the original 1903 edition and into the period shortly before he began preparing the major 1909 revision, and then into his retirement and the period in which he was producing his important philosophical series, The Background Books: The Philosophy of the Holy Earth.

The various sections in this edition need not be read in any particular order. Different sections will variously appeal to different readers. Many may want to skip right to Bailey's text, at least the first time through. Educators will be sure to read Dilafruz Williams's essay. Those who want the story behind the book and a bit of Bailey's biography will enjoy "It Is Spirit." For the general reader, however, the book is arranged in an order intended to be as useful as possible from front to back. In the same manner, Bailey originally organized *The Nature-Study Idea* to be readable either from front to back or out of order, according to the reader's interest—Part I walks through his nature-study philosophy systematically (with plenty of entertaining anecdotes along the way), Part II focuses more on the teacher's point of view and the "outlook" underlying the philosophy, and Part III applies experience to a variety of practical questions and concerns raised by Bailey's readers. His text is worth reading through in its entirety, but it also invites rereading, skipping back and forth, and cross-referencing. Indeed, Bailey scatters parenthetical cross-references throughout the book, regularly directing readers to other pages where similar ideas are discussed.

It is our hope, in launching the Liberty Hyde Bailey Library with this new edition of one of Bailey's most foundational and influential works of popular philosophy, that we both help to introduce Bailey's writings to a wider contemporary audience and bring a wealth of resources together for renewed academic appraisals of Bailey's work in the realms of educational and environmental philosophy and literature. It is the opinion of the series board that Bailey's work speaks just as powerfully today as it did a century ago, that in fact we need the clarity of his words as an inspiration and guide through the complexities of our current moment, and that this book ought to find a treasured place on the bookshelf of every lover of the open country.

BRINGING EDUCATION TO LIFE AND LIFE TO EDUCATION

Contemporary Relevance of Bailey's Nature-Study

DILAFRUZ R. WILLIAMS

Nature-study is coming more and more to be an out-of-door subject,
for the child's interest should center more in the natural
and indigenous than in the formal and traditional [. . .]. There
can be no effective sustained nature-study when the work is
confined in a building.

Liberty Hyde Bailey, "The Common Schools and the Farm-Youth,"
1907 (see Related Writings, this volume)

Over a century ago, Liberty Hyde Bailey resisted the then-emergent and
tantalizing trappings of the industrial revolution that uprooted rural
communities with the lure of urban life in the name of progress. He saw
an increasing disconnection of children from place, community, land,
soil, and nature at a time when science teaching was just emerging in the
lower grades and was becoming an indoor activity mediated and con-
trolled by lifeless models, stuffed specimens, and books to the detriment
of direct bodily and hands-on experiences with and in nature outdoors.
Equally disconcerting for Bailey was the growth of lifeless school and
classroom structures. Believing firmly in developing the child's outlook

on the world through exploratory and unfiltered outdoor experiences, he promoted values that were "natural" and "indigenous." This was in direct contrast with rote memorization and book learning "confined in a building."

Bailey's relevance for our times should not be underestimated. His warnings resonate today with educational practitioners and policy makers alike. Close to fifty million children and youth, prekindergarten to grade twelve, are enrolled in public schools in the United States, spending about ten months in school each year. No matter the age of the child, much of this time is expended within the confines of the concrete walls of classrooms and school buildings that are surrounded by asphalt and blacktop, parking lots, grounds that are chemicalized to curb weeds, or artificial turf for sports. The modern mind-set that values efficiency also drives the ever-expanding scale of school buildings.

More and more, our formal organizational structures and institutions devoted to educating children, youth, and adults are characterized by human-made, lifeless built environments. Guided by the Cartesian mechanistic paradigm that ignores the rhythms of natural cycles, today's common pedagogical accessories and milieus include science and media laboratories, books, and technological gadgets such as smartphones, laptops, computers, televisions, screens, and projectors. Via the internet, children are growing up more connected with distant others, elsewhere, than with their own locale and place. Bailey's fears of the disconnectedness of education from nature are playing out over a hundred years later, with children who can often recognize dozens of corporate logos but cannot identify a tree on their own school grounds or in their neighborhood.

The cycle of daily news reporting the dramatic environmental degradation that affects our life systems is indeed sobering and daunting. We are familiar with environmental threats to humanity, the urgency of climate change, and the sociopolitical and economic havoc resulting from the ongoing pillaging of Earth's life-supporting gifts—her soils, water, wildlife, air, and natural resources. As with Bailey's perceptions, we find critics of modern education deeply concerned about how schooling aggravates this problem. Hence, for several decades, countless grassroots efforts have emerged across rural, urban, and suburban schools and their communities, challenging the educational status quo to change its trajectory toward building hopeful and life-sustaining relationships with the natural world.

Hopeful Signs and Promising Possibilities

In *Blessed Unrest*, Paul Hawken provided insight into a movement in which thousands of diverse organizations and communities, large and small, were involved in restoring the environment and fostering social justice across the globe. This movement, according to Hawken, had no specific name, leader, or location. Similarly, efforts to bring life to education and education to life in the United States have been ongoing, without the overarching narrative of a megasolution.

Since the 1990s, there has been growth and a surge of interest in a variety of programs that have embraced active engagement of children and youth outdoors, along with the restructuring of some schools to advance place-based education. Countering standardized curricula and high-stakes standardized tests promoted in the 1980s to advance American domination globally, these programs can be broadly clustered as nature-based education, garden-based education, environmental education, outdoor education, conservation education, sustainability education, earth education, adventure education, indigenous education, social-justice education, climate change education, watershed education, and more.[1] Furthermore, the pandemic in 2020 opened up opportunities for outdoor education at school sites, with school grounds and school gardens serving as safer havens for learning when indoor classroom activities were restricted or disallowed.

Starting as grassroots initiatives, the long-term growth of these programs is often dependent on fostered partnerships and the availability of financial and human resources. The Children and Nature Network (https://www.childrenandnature.org/) and the North American Association for Environmental Education (https://naaee.org/) are two large networks that have championed research, professional development of teachers and educational leaders, and advocacy with policy makers. These organizations have raised the profile and increased the legitimacy of the outdoors and nature connections in education.

Too numerous to recount here, perhaps a glimpse into one type of nature-study, school garden programs, can provide insight into practices that foster life experiences on school grounds, delivering the needed understanding and relationship with nature urged by Bailey. Some examples of programs that have stayed the course are the Edible Schoolyard

Project, Farm to School Garden Networks, Green Schoolyards, Growing Gardens, the Learning Gardens, Living Classroom, and KidsGardening (website links for these and a wider sample of the prominent garden-based programs and initiatives are provided below this essay). This list is by no means exhaustive—every state has multiple networks of gardens founded by either school communities or not-for-profit organizations to enhance students' connections with food, healthy eating, social-emotional development, and academic learning.

Garden-based education, which embraces many of the tenets of educating for, in, with, from, and about the environment and nature, appeals to a wide range of constituencies across urban–suburban–rural landscapes. Students from a variety of backgrounds, including low-income and culturally diverse students, build gardens on school grounds and grow and harvest food while simultaneously learning various subjects outdoors with hands-on experiences in the garden. Increasingly, indigenous knowledge and traditional practices are honored and integrated in garden curricula. In these programs, science tends to be the most integrated subject, but language arts, poetry, mathematics, and social studies are also taught.

In the Learning Gardens program in Portland, Oregon, for instance, books and curricula come to life with multiple interdependent feedback loops for holistic education. Students are awakened to the concept of seasons linked with growing food. Similar to the Edible Schoolyards program in Berkeley, California, and the Manzo Gardens program in Tucson, Arizona, in the Learning Gardens, students become aware of how the health of humans, the health of the land, and the health of their communities are intertwined. Learning in and with nature provides tangible, pragmatic, and embodied meaning.

Integrating various subjects and experiences, garden programs often advance student motivation and engagement. General pedagogical principles and insights emerge across the learning gardens and are elaborated through the use of an acronym—GARDENS—in Williams and Brown's *Learning Gardens and Sustainability Education* (2012):

1. Cultivating a Sense of Place—Groundedness: In our globalized era, garden-based nature education cultivates a "sense of place." Just as Bailey argued in the past, modern students cannot deeply learn

about place solely from a textbook. Gardens depend on climate, soil, terrain; there are infinite variations for design and use, thereby giving students a fruitful and practical sensibility to grow and cultivate their own sense of place. Individual gardens are finely tuned local expressions of natural phenomena like sun, rain, wind, air, and more. However, students must also be engaged critically to address issues related to food justice, ecological justice, voice, and inclusion. As one middle school student stated: "Food that we grow in learning gardens gives us *skills* so that we can learn to become independent and not always rely on large businesses to feed us. I think about the hungry and the homeless and how I can be of service to them."

2. Fostering Curiosity and Wonder—**A**we: Children's latent capacity for curiosity and wonder must be nurtured. We have all heard young children and youth ask penetrating questions about life-giving organisms, wondering about things ranging from the wriggling of worms to eggs in the chicken coop to dewdrops on leaves. Garden settings provoke endless marvel, wonder, and questions. For instance, it is not unusual to find students to be fascinated by organisms they discover in the compost built as part of the garden. Students of all ages tend to pause to examine these creatures when they realize that soil is alive and to learn about the nuances of decay, death, birth, and life that affect soil texture, smell, porosity, and color. An important question is explored: How does soil become part of the recursive food web?

3. Discovering **R**hythm, Scale, Patterns: Together, rhythm and scale describe patterns of relationships, and those relationships form the conceptual basis of an ecological model. Understanding relationships as central features of living systems shifts attention from the parts to the whole. We are surrounded by natural rhythms that frame life: the daily rising and setting of the sun and moon, the turning of the planet, the changing of the seasons. In school gardens, students discover that relationships among natural rhythms at different geographical and biological scales create an endless diversity of ecosystems and niches in time and space, all worthy of inquiry.

4. Valuing Biocultural Diversity: Biological and cultural diversity are inextricably linked. In the gardens, students learn about the dynamic and reciprocal interdependencies among human activities, cultures, and environments. Educational gardens have been excellent sites for initiating students into lifeways that are respectful of biocultural diversity and environmental justice because such gardens are uniquely situated at the symbolic and literal crossroads of biology and culture.

5. Embracing Practical Experience: It is widely recognized that not all students flourish in learning environments that are didactic, abstract, and focused on reading and writing—something Bailey lamented about, too. Many students learn best through practice or bodily engagement. Through planning and planting a school garden, students are provided multiple opportunities to engage diverse ways of learning and are brought into conversations about life. Experiences with the real world teach students in profound ways.

6. Nurturing INterconnectedness: Appreciation of the interconnectedness of nature as one of its defining characteristics, along with the realization that we too are inherently interdependent upon one another and with nature, can significantly affect our interactions and relationships with others, including the biotic world. In educational gardens, students use their hands and their hearts to intuitively learn in the sun and the soil, which leads to a fuller bodily understanding of relationships and interdependence.

7. Awakening the Senses: Sensory engagement allows students to make meaning in ways that go beyond a particular school subject. For instance, at an elementary school, students learned to grow food and to harvest underground roots—beets, carrots, turnips. While washing, cutting, and tasting were part of the ritual, they also wrote haiku. An example shows the depth of learning taking place:

Beet's squashy bursting with blood,
Red dye everywhere
Smells like buckets of dirt.
Over-sugared. Makes me gag.

A school's vitality emerges with its numerous partnerships and networks that include parents, community members, neighborhood associations, not-for-profit organizations, governmental institutions, philanthropic foundations, businesses, and universities and colleges, among others. An added and equally important benefit of connecting education with nature is civic engagement and participation by diverse communities for a vibrant democracy, as both Bailey (*Nature-Study Idea*, 1903) and John Dewey (*Democracy and Education*, 1916) recognized. On April 7, 2022, the bipartisan No Child Left Inside (NCLI) Act was introduced in the 117th Congress of the United States (H.R. 7486/S. 4041; 2021–2022). In an earlier NCLI Act, Congress acknowledged the need for promoting the out-of-doors and real-life experiences with nature to mitigate our contemporary schooling's separation from life (H.R. 3036/S.1775; 110th Cong., 2007–2008). Several states have also passed their own NCLI Acts. While Bailey would find this validation promising, these nature-based initiatives won't be sustainable without much-needed investments to appoint designated school personnel who can be built into educational budgets.

Congruent with innumerable initiatives to promote nature connection with education, syntheses and meta-analyses of research studies published since the late twentieth century provide evidence for a multitude of benefits and positive outcomes of educational efforts that connect students from preschool to high school with nature and the outdoors.[2] These outcomes fall under three broad and overlapping categories: academic outcomes; social, emotional, and behavioral outcomes; and health-related outcomes. Natural environments support attentional functioning, a sense of curiosity and wonder, multisensory learning, food literacy and healthy eating habits, physical activity, environmental stewardship, school bonding, community/parental involvement and intergenerational learning, motivational engagement, social and moral development, and vocational skills.

Bailey clearly recognized—as we also find in children and youth directly engaged with nature—that knowledge and understanding of nature content are linked with emotional attachments. "The keynote of nature-study is to develop sympathy with one's environment and an understanding of it," wrote Bailey (The Common Schools, in Related Writings, this volume). Emotive engagement is as important as intellectual understanding. Indigenous traditions and cultures have valued this engagement for millennia. The 4-H movement that Bailey helped to inspire values holistic learning and honors the "head, hand, heart, and health" in nature education. *Place* can be a nexus for critical outdoor learning, where the city, the watershed, the farm, or the community become a classroom to interrogate issues of social justice, eco-justice, and food justice. Bailey's hope for kinship and sympathy developing from nature-study can and should emerge at this nexus.

During these dire times of environmental degradation and climate change, love, not fear, will motivate humans to act.[3] Hope continues to sprout because the place-based practical pathways to human agency and empowerment advised by Bailey are possible. Bailey's pragmatic philosophy that "teaching begins with the actual, the tangible, the significant" (The Common Schools, in Related Writings, this volume) is also in tune with ideas currently being put forth by certain Indigenous educators and theorists. Botanist Robin Wall Kimmerer, a citizen of the Potawatomi Nation, reminds us that traditional ecological knowledge is "born of long intimacy and attentiveness to a homeland and can arise when people are materially and spiritually integrated with their landscape."[4] Bailey encourages educators to engage students with nature and to decompartmentalize schooling, as nature's gifts make us wonder: where does science end and poetry begin? The softness of a rose petal, the warmth of cooked compost, the sweetness of an apple, the aroma of wet soil, the tweet of a warbler, and the spectacular fluttering of a hummingbird must be experienced. Nature-study can serve as a significant leverage point in this endeavor.

Program Links

Children and Nature Network: https://www.childrenandnature.org/
Common Roots: https://commonrootsfarm.org/
The Edible Schoolyard Project: https://edibleschoolyard.org/
Farm to School Garden Networks: https://www.farmtoschool.org/
Forest School Education: https://www.forestschools.com/
Green Schoolyards: https://www.greenschoolyards.org/
Growing Gardens: http://www.growing-gardens.org/
KidsGardening: https://kidsgardening.org/
The Learning Gardens: https://learning-gardens.org/
Learning in Places: http://learninginplaces.org/

Life Lab: https://lifelab.org/

Living Classroom: https://www.living-classroom.org/

Manzo Elementary School Garden: https://www.gomanzo.com/

Native American Gardening: https://www7.nau.edu/itep/main/SGardn/ Resources/res_nativeam

North American Association for Environmental Education: https:// naaee.org/

Veggielution: https://veggielution.org/

"It Is Spirit"

The Genesis of The Nature-Study Idea

John Linstrom

In history, it is an oft-repeated story that the success of a movement
for the betterment of the world has been dependent on the genius
of the man who first had it in charge. It was[,] therefore, fortunate
for the Nature Study movement that in 1897, one year after the first
appropriation was made for the work by [New York] State, Professor
Roberts placed the whole enterprise in the hands of Professor Liberty
Hyde Bailey, at that time head of the Department of Horticulture in
Cornell University. NO wiser step could have been taken. Professor
Bailey is a great man by any standpoint, but perhaps his greatness
is never more in evidence than in his genius for leadership. He had
great vision concerning this Nature Study movement, and great faith,
also. He was especially fitted for the work, for he had been born and
had spent his childhood on a farm, and had, as gifts from birth, an
innate love of nature in all of its moods and the poet's imagination
that gave him vision beyond his horizon.

Anna Botsford Comstock, *The Comstocks of Cornell*, pp. 225–226

It would be well if *The Nature-Study Idea* were in the hands of every
person who favors nature-study in the public schools, of every
one opposed to it, and, most important, of every one who teaches
it or thinks he does.

The Tribune Farmer, circa 1903, as quoted
in Macmillan advertising material

The Schoolteacher Who Lit the Spark: Julia Field

In *The Nature-Study Idea*, Liberty Hyde Bailey claims that he was motivated to compile a book on "some of the more salient features" of the
nature-study movement by the "common [. . .] misconception of the
meaning and mission" of that movement.[1] His "main thesis" to answer
that misconception, he went on to write, was "that nature-study teaching
is one thing and that science-teaching for science's sake is another," and
on that point, he wrote, "I have no hesitation" (Part I, Chapter I; hereafter, I.I). It was a curious argument for a professor of horticultural science
to make, but it essentially amounted to a defense of the movement he had
been observing among working teachers, many of them women, against
a series of prominent attacks that had been leveled against them by university scientists, most of them men. While contemporary criticisms of
nature-study seem to hover in the background throughout Bailey's book,
he wastes relatively little time addressing either them or the critics themselves, preferring to focus on the positive philosophical ideals of the actual
educators he had been supporting from his place in the College of Agriculture at Cornell University. And while the term "nature-study" was still
relatively new, Bailey understood that the idea had been around longer—
when he, along with his colleagues Anna Botsford Comstock, John Spencer, and others, began to take stock of the movement as it was manifesting
across the state of New York, he found a validation of many of the ideas
that had informed his thinking on education and the ways of knowing nature since he was a young boy.

The story of *The Nature-Study Idea*, therefore, begins much earlier.
We might start sometime around the year 1870, when a thin, wiry farm
boy with an unusual name came to school with an unusual request.
Young Liberty had walked a mile through the woods to get to the village
schoolhouse, a simple four-room structure that was named Central School
because it happened to be the one located in town and that was "called
a high school, because one room was over the other," as the pupil would
recall in a speech many years later.[2] He made this walk every school day
during the short periods that classes were in session in that farming community, but this day he came bearing an unusual textbook, one that he
had picked up in the cabin of a fellow settler somewhere in or around
the village. Books were scarce in the boy's home, but he devoured them

Figure 1. Central School, South Haven, Michigan. This is the four-room building as it was built in 1858, the year Bailey was born. A remnant of the ravine Bailey remembered, west of the school, survives on the old school grounds and is known as Baer Park, although the property has been purchased by a private condominium developer and its future is uncertain. Courtesy of the Richard Appleyard Collection of the Historical Association of South Haven.

whenever he could lay his hands on them. This tattered old book was missing its cover board and title page, so he never learned the name or author, but it was a textbook on natural history. At that time, the "three Rs" of reading, writing, and arithmetic still ruled the country schoolhouse, and the sciences had little place. The law did provide each student the right to request one additional subject, although teachers in turn reserved the right to refuse any such topic that they felt unqualified to teach. Liberty loved the outdoors, and from a young age had spent countless hours exploring the woods around his family's farm, observing the creatures that made their home in the little brook and collecting plants and animals to raise at home. His favorite part of school might have been the ravine that bordered the small frame building, which some sixty years later he described as, "[w]ith one exception, the most wonderful place in the world," and which to his child's mind "was hundreds of feet deep. It was inhabited by monsters, and wonderful creatures. Had we seen them? No; that was why we knew they were there."[3] When he approached his teacher Ms. Field that day, tattered textbook in hand, with the request for a course on natural history, he likely did so with no small amount of desperate hope.

"Why Liberty," Bailey would remember his teacher telling him, when he recounted the story years later, "I should like that very much. I know nothing about natural history. You may recite to me every day from this book." No one else was interested, so it became a class of two: teacher and student. To take up a subject so far out of her comfort zone might have been enough, but Ms. Field didn't stop with mere recitation, which was then the most common form of instruction in schools like hers. Learning along with her pupil as he recited day by day, she would also ask him to report on the animals and plants he had seen in his walks through the woods or his work in the fields. "She claimed to know little about [the subject,]" Bailey would recall, "but she challenged every observation I made. I knew the birds and the animals, and every morning I would report what I had seen and she would listen, and would ask questions. It was the best teaching possible."[4] She pushed him to make the learning a part of his everyday life.

The most memorable experience from that informal course in natural history came one day when Ms. Field said, "Liberty, I am very sorry for you."

"Why should anybody be sorry for me?" Bailey asked.

"You're going through a beautiful world with your eyes shut. You see nothing."

"Why, no," he would remember objecting: "I am not blind. I see everything. I see all the birds and the trees." The teacher may have lifted her eyebrows to this remark.

"You come to school every day bringing your dinner?" she pressed.

"Yes."

"Where is your walk?"

He told her that he came through the woods.

"How many maple trees are there along that walk?" she asked him.

The boy could not answer that day, but by the next day he could, and he gave the number with a good bit of pride. "Yes," she answered him; "how high are the tops from the ground on the different maple trees?"

Again, he didn't know, but he did by the next day—and so the exploration continued, observation by observation. Bailey would recall that when she didn't know the answer to a question she herself asked, she would simply say, "I don't know; I want you to tell me." And in that way, he recounted many years later, "she kept me growing for a blessed year."[5] In a different interview, he recalled, "It was the first and greatest lesson I ever had in natural history [. . .]. That teacher had the magic touch of inspiring the desire for light in a young man."[6] That inspiration would grow into a lifelong dedication to close observation and memory of the world he inhabited.

"Fundamentally," he would famously declare, "nature-study is seeing what one looks at and drawing proper conclusions from what one sees; and thereby the learner comes into personal relation with the object." And, "as with all education, its central purpose is to make the individual happy" (I.III).

Julia Field, a classically educated woman of about thirty from northern Illinois who had settled in the small frontier town and turned to teaching to support herself, could not then have known that her pupil's unusual name would become one of the most familiar and influential names in the educational movement to teach children through firsthand, natural experience rather than through mere recitation from books.[7] In a way, there in the frontier village of South Haven, she lit the spark that would help to ignite the nature-study movement and bring it into international attention. That movement, which emerged right as teachers and pedagogues around the country were struggling with the question of how best to integrate the sciences into elementary curricula, left an enduring legacy of hands-on, child-centered, experiential learning in science education, and it also spawned innumerable extracurricular clubs, camps, and museum programs to take children outdoors. Even the modern concept of the "field trip" owes its genesis, in part, to the nature-study movement.[8] In 1909, when Bailey thoroughly revised *The Nature-Study Idea*, his influential manifesto on the subject that had already gained notoriety and gone through multiple reprintings since its first appearance in 1903, he dedicated the new edition of his landmark book to the childhood teacher who had, perhaps without knowing it, helped to change the face of American education. Julia Field didn't need to be a scientist to spark this lifelong sympathy with nature and love of direct investigation—the critical thing was that she took her lessons outdoors, and she knew how to inspire a sense of wonder, a *spirit* of inquiry, that would drive her pupil's curiosity for a lifetime. Bailey memorialized her in his dedication simply as "a teacher who allowed a boy to grow." By the time he was pulling his book together, the idea that nature-study should be taken out of the hands of such teachers and given over to university scientists must have struck Bailey as preposterous. As he knew well, "[t]he power that moves the world is the power of the teacher" (II.I).

The Idea in Practice: Bailey the Teacher

"The nature-study movement is the outgrowth of an effort to put the child into contact and sympathy with its own life," Bailey stated in opening a speech to the National Education Association in 1903.[9] It was a movement of ideals, in other words, and not tied to any one discipline or field of study. He also insisted that it was a grassroots movement originating in the common or public schools, and not in the universities. He knew this because he had seen it—when Cornell University, where he was professor of horticulture, received funding to investigate the potential to serve the state's agricultural interests by supporting nature-study in the schools, he was personally involved in traveling on foot and by horse and buggy with his nature-study colleague Anna Botsford Comstock from school to school around the countryside, meeting with teachers who were experimenting with child-centered outdoor learning.[10] He had experienced a similar kind of education under Julia Field's tutelage as a child, and Field wasn't working from any plan or movement but just out of a gifted teacher's intuition. But Bailey had not only seen nature-study at work: what is less often appreciated is that he had been a nature-study teacher himself, following in Field's footsteps, in the years before the movement had a name.

When Bailey left the farm to pursue a degree in horticulture at Michigan's fledgling State Agricultural College in East Lansing (later known as Michigan Agricultural College and eventually as Michigan State University), he had to find work to help support his studies. The fall term at the college was kept to just three months at that time to enable students to teach at district schools during the long winter term, and Bailey took advantage of the opportunity.[11] He secured a job at the rural Carl School in the winter of 1879–1880, at the age of twenty-three (he had entered college at twenty-one[12]), and while teaching there he boarded at a farm across the road. He may have been green as a teacher, but one of his pupils, William Donley, who would spend most of his life in the Carl School District and would go on to sit on the school board for over five decades, would later remember Mr. Bailey as one of the best teachers he ever had.

The school apparently had a reputation for rough children, and many teachers didn't last long if they couldn't stand up to the rougher boys. Bailey had known such a situation as a student at Central School in South Haven, and he was prepared, although many had their doubts at first about the "spindly fellow." The reputation he gained wasn't one of a tough teacher, though—quite the opposite, Donley would later report: "[H]e had the record that he never punished a scholar. That was quite a record for this district." Corporal punishment was still the norm, so Bailey's approach was notable. As he would write over two decades later in *The Nature-Study Idea,* "The greater number of mischievous and refractory children can be interested in some piece of personal work or investigation. The boy who is 'licked' at home and punished at school is likely to spend his time midway between the two; and yet he may be easy to reach if only he is understood" (III). Most male teachers who came to the school would end up spending recess periods wrestling the more "refractory" boys, but not so with Bailey, who instead would sit on one of the desktops in the school, hold up a twig that he had collected from a tree or bush outside, and tell his students what he knew about it—and, apparently, he held a crowd. But he did not simply lecture indoors about outdoor topics. Like Mrs. Field, he sent his students out. In 1951, over seventy years after the events took place, William Donley would fondly recall the following story, as reported in a magazine feature that year:

"One day, [Bailey] told three of us to go down to the marsh and cut some pussywillows with all the catkins out," Mr. Donley said. "In those days, teachers used willow switches to whip a student who got into trouble. The three of us couldn't figure out who was in for it. While we were cutting those pussywillows, we asked each other who'd gotten into mischief."

After they'd each gathered a big armful of the willows, the boys took them up to the schoolhouse. Dr. Bailey was sitting at his desk, a big handmade affair, eight feet long. He was studying, as usual, but he looked up, thanked the boys, and told them to put the willows under the desk.

"Then we began to tumble," said Mr. Donley.

"Mr. Bailey said 'Boys, I'd like to show you something about those pussywillows. Tomorrow's Saturday, but if you'd like to come in, I'll be here, and we can see what we can learn about willows.'

"We were there, and some of the other boys from the school were with us. He showed us quite a bit about those willows: how the catkins are shaped, how the bud is fastened to the stem, and how the twigs grow. Did you know that every third catkin is in a line right up the stem?"

Dr. Bailey and his students studied the pussywillows for about a week. Then they examined thorn bushes, the crabapple and the wild plum.

"After that," Mr. Donley said, "he asked us if we had any questions about the bushes and small plants growing on our farms. He showed us things we'd never thought to look at before.

"All the while he was teaching us, he was learning too. He was a good teacher and he made us want to learn."[13]

The Philosopher Gathers His Manuscript

In his adult life, Liberty Hyde Bailey was as much a social theorist and philosopher as he was a scientist, as much a popular author and journalist as he was an academic, and in each of these roles he remained an effective and impassioned teacher. He considered education to be the foundational soil out of which social progress would necessarily grow, and, as *The Nature-Study Idea* makes clear, he maintained an optimistic faith that public school teachers, properly equipped and supported by their communities, were precisely the leaders who would help to build a better world through their teaching.[14]

The nature-study idea, broadly understood, applied to all students, whether in rural or urban areas, and the movement was strong in cities like New York and Chicago—Cornell's program, in fact, worked directly with teachers in New York City and instructors at the Teachers College at Columbia.[15] At Cornell, however, the work was originally taken up in order to address demographic changes that were concerning rural leaders, as more and more people were forced out of farming by the economic depression of the 1890s and consequently moved to towns and cities.[16]

Bailey knew that the problems were not merely economic but also cultural: rural people had for a long time been made to feel like second-class citizens, and children inherited the sense that success was to be found in the cities and not on the land. Cities were growing in importance as centers of capital, and while the colleges of agriculture might help improve yield and make farming somewhat more lucrative, Bailey knew that these children needed more than a monetary argument for remaining and investing in their rural communities when they grew up:

> In a certain rural school in New York state of say forty-five pupils, I asked all those children that lived on farms to raise their hands: all hands but one went up. I then asked all those who wanted to live on the farm to raise their hands: only that one hand went up. Now, these children were too young to feel the appeal of more bushels of potatoes or more pounds of wool, yet they had this early formed their dislike of the farm. Some of this dislike is probably only an ill-defined desire for a mere change, such as one finds in all occupations, but I am convinced that the larger part of it was a genuine dissatisfaction with farm life. These children felt that their lot was less attractive than that of other children; I concluded that a flower-garden and a pleasant yard would do more to content them with living on the farm than ten more bushels of wheat to the acre. Of course, it is the greater and better yield that will enable the farmer to supply these amenities; but at the same time it must be remembered that the increased yield does not itself awaken a desire for them. I should make farm life interesting before I make it profitable. (I.VII)

Through nature-study, teachers had the power to open children's eyes to the wonders of the natural world that surrounded them in the country, much like Ms. Field had helped to do for Bailey in his youth, and he hoped the love of the land might motivate a generational reinvestment in country life. Over time, experience seemed to bear this out: "One of the most significant comments I have heard on nature-study work came from a country teacher who said that because she had taught it, her pupils were no longer ashamed of being farmers' children" (I.VII). But Bailey also knew that this principle extended just as well to urban and suburban children—"What can be done for the country child can be done, in a different sphere, for the city child" (I.VII). Endless curiosity, fueled by the training in scientific observation that nature-study also enabled, could provide intellectual pleasure and love of life in any setting. As he would wryly observe, "The happiness of the ignorant man is largely of physical pleasures; that of the educated man is of intellectual pleasures. One may find comradeship in a groggery, the other may find it in a dandelion; and inasmuch as there are more dandelions than groggeries (in most communities), the educated man has the greater chance of happiness" (I.III).

By the time the book was published in 1903, the nature-study program that Bailey helped establish at Cornell University had been doing its work for seven years to resounding success far beyond the state of New York. He and his colleagues in Cornell's Nature-Study Bureau, including Anna Botsford Comstock, "Uncle John" Spencer, and Alice McCloskey, were in high demand at teachers' institutes; their summer courses were full; and their various series of nature-study leaflets for teachers and students were officially being circulated all around New York State and unofficially circulated around the world—in 1904, twenty thousand leaflets and eighteen thousand bulletins were printed to meet the demand, and by 1908, Bailey reported that they were publishing seven thousand regular leaflets and thirty-seven thousand supplementary leaflets for children *per month*.[17]

As teachers and administrators sought engaging ways to incorporate the burgeoning scientific disciplines into their curricula, nature-study promised to quicken children's interest in the sciences by building on what many educational reformers believed to be the relatively universal interest of children in the world around them. More important for Bailey, however, was the movement's promise to equip children—and, really, anyone who caught the spirit of the movement—to live happier, fuller lives. He bristled at accusations from some of his fellow university scientists that nature-study teachers lacked scientific rigor and that the open-ended learning they modeled might lead to an incomplete understanding of natural science. For Bailey, the point wasn't to make more scientists but to enrich the lives of as many students as possible through careful and scientifically informed encounters with the more-than-human world, acknowledging that most of those children would never go on to the academic pursuit of science and become a specialist. *The Nature-Study Idea* emerged from the lectures he had been delivering to teachers, and he directed the book toward teachers concerned with teaching, not toward academic scientists concerned with the perpetuation of their various disciplines.[18] It was informed, moreover, both by his own experience as a teacher and by his extensive contacts with local teachers in upstate New York (see, for instance, the frontispiece to this book) and with faculty at normal schools (what today we might describe as teachers' colleges).[19]

By the time Bailey began to conceive of the book, he had been steadily engaged in the work for half of a decade, but it wasn't his first foray into educational writing. His very first book in 1885, *Talks Afield about Plants and the Science of Plants*, exhibits some of his later nature-study philosophy, and it even reproduces the lesson he would attribute to Julia Field using nearly the same words, in the chapter "The Importance of Seeing Correctly":

> It is surprising that many of the commonest and most interesting of everyday phenomena, though they lie right before the eyes of every man, are never seen by the great majority of the people. Most persons are walking through a wonderland with their eyes shut. The interesting things detailed in these pages are but a very few random leaves rudely torn from the book of nature. The leaves that remain are fully as inviting, and they are doubly profitable when Nature herself tells the story.[20]

In the ensuing decades, he became known for his highly successful books on gardening and horticulture, including the monumental *Cyclopedia of American Horticulture*, which began appearing in 1900, and for several textbooks geared toward college-level agricultural coursework. But in addition to all of these, he had also already authored two textbooks meant to support nature-study work in the public schools. Written for teachers in elementary and secondary schools, *Lessons with Plants* (1897) had been his first foray into elementary textbook writing, a richly illustrated volume that outlined lessons based on the simple, sustained observation of plants in their natural habitats. The lessons there on the observation of simple twigs recall the pussywillow observations that William Donley remembered from Bailey's early teaching at Carl School. Then, in 1900, Macmillan published Bailey's *Botany: An Elementary Text for Schools*, which was written for students and advertised as a companion to *Lessons with Plants*. The nature-study work under Bailey's leadership at Cornell was by then accelerating and proving successful, and it was with the added success of these textbooks that, the following year, Bailey approached Macmillan with a third sort of nature-study book—one that would go beyond mere sample lessons and get at the philosophical heart of the movement that was underway.

Bailey was primarily known to the editors at Macmillan at that point as a prolific and effective author and editor of textbooks and reference works—not as a philosopher or litterateur. Yet he had always been interested in more literary and philosophical writing. One of his earliest complete book manuscripts, written in the winter of 1886–1887, was a literary travel narrative based on an expedition in the Boundary Waters of northern Minnesota and southern Canada with Anishinaabe guides; it was titled *Onamanni* after the name given by the Bois Forte Band of Ojibwe to one of the lakes where he began and ended the botanical expedition. He revisited that early manuscript in 1899 and wrote a new brief foreword for it, apparently with renewed hopes of finding a publisher.[21] The text had afforded ample opportunities for Bailey to try out some of his more philosophical ideas about the human place in and responsibility toward the natural world. While that manuscript has been preserved, it is not clear whether he did follow through with submitting it, or to whom. Also in 1899, he completed an apparently new literary manuscript, titled *Beside the Still Waters*, that he sent to Macmillan editor George P. Brett. He framed the manuscript as a response to Brett's suggestion that Bailey write "a story which should have a country-life setting" and asked Brett for "an opinion as to its value." Brett seems to have read it through himself and offered to find a time to meet with Bailey in New York City to discuss possibilities for rewriting and strengthening it—an opportunity Bailey indicated he would be glad for, though "not with a view to rewriting the MS. but as a matter of general interest and education to myself."[22] That book also never came to be, and the manuscript is not known to exist.

A year and a half later, in January 1901, Bailey again wrote to Brett, recalling his attention to "a brief correspondence [that we had] last fall in regard to a book which I was then preparing on 'The Nature-Study Idea'. You asked me to send you manuscript, but I replied that I should prefer to wait a short time. I have waited longer th[a]n expected in order that my ideas on some of the points should mature. I am now sending you the manuscript, however, by express." His aspirations for the book show somewhat in his vision for the printing of it: "It was my own thought if the book is published at all, that it might be made a 16 Mo [sextodecimo, meaning 5 × 7.5 inches] with untrimmed edges and put up perhaps in an artistic way."[23]

There was no immediate response from Macmillan, and the silence was such that Bailey wrote again on February 22 asking for a decision. It would be a month yet before the rejection came. While the letter from Brett has not been found, and that first draft of the manuscript is not known to exist, Bailey's response on April 3 indicates that he had not yet lost faith in it. He writes, "The delay in looking over the manuscript has been of no inconvenience to me, and your estimate of it is no surprise and very little disappointment. I am inclined to think, however, that you underestimate the interest in the nature-study movement. However, I have no desire whatever to press the subject and am willing to let the matter drop, so far as we are concerned."[24] It seems that he thought even then that he might find another interested publisher.

It so happened that at that very time Bailey was in the process of launching an ambitious new magazine with Doubleday, Page and Company titled *Country Life in America*, which would first appear in November of that year and would prove to be another early testing ground for Bailey's nature-study philosophy. He had an active hand in the magazine's contents, penning many editorials and brief articles himself, particularly to fill out the early issues, and a number of his nature-study colleagues contributed. Anna Botsford Comstock even came on as poetry editor and would later recall that "Bailey had two things clearly in mind" for the magazine: "one, to give practical help to farming and horticulture; and the other to lead people to appreciate the beautiful in nature and learn the many things of interest in the fields and woods."[25] Toward the latter goal, his editorials and the magazine content of the early issues generally show that, between the essays on farm and garden methods and appreciations of rural life, he was also interested in using the magazine as one more vehicle for articulating the Cornell program's nature-study philosophy. A number of these editorials were destined to appear later in *The Nature-Study Idea*, but whether Bailey had written them originally for the rejected 1901 manuscript or for the magazine itself is unclear. What is clear is that he became deeply occupied with honing the kernel of this "idea" and getting it down in print.

The magazine's second issue, in December of 1901, opened with Bailey's article "An Outlook on Winter," which beautifully melded an appreciation of the winter months with his own childhood experiences learning from nature's book in the woods and streams surrounding his home on the Michigan farm. "The lesson is that our interest in the out-of-doors should be a perennial current that overflows from a fountain that lies deep within us. This interest is colored and modified by every passing season, but fundamentally it is beyond time and place. Winter or no winter, it matters not: the fields lie beyond" (II.VII). Two months later, for his monthly editorial in that February issue, he tried out the title of his rejected book manuscript, "The Nature-Study Idea," and in two full pages of the large-format magazine attempted to boil down the *idea* of this new movement to bring students outdoors, an effort he believed went much deeper than subject matter or pedagogy. "Of late years there has been a rapidly growing feeling that we must live closer to nature," he begins, "and we must perforce begin with the child. There is an effort to teach this nature-love in homes and schools, and the subject is called nature-study. It would be better if it could be called nature-sympathy."[26] From that point he proceeds to try out many of the ideas that would eventually appear in the book; indeed, much of the particular language and phrasing of the brief editorial was either lifted from the manuscript or destined to be carried over into a later draft. You can hear him trying out the wry statement quoted previously: "As with all education, [nature-study's] central purpose is to make the individual happy; for happiness is pleasant thinking. The happiness of the ignorant man is largely the thoughts born of physical pleasures; that of the educated man is the thoughts born of intellectual pleasures. One way to lessen evil-doing is to interest the coming generation in dandelions."[27] A bit less colorful than the "groggery" image he would later land on, but the sentiment—and the dandelions—are there.

April of 1902 included an editorial titled "The New Hunting," in which Bailey articulated a nuanced view of the motivations behind hunting for sport, which he believed at their best centered around a desire for deeper engagement with the natural world, and suggested that, as many people in industrializing America depended less and less on hunting for sustenance, other forms of enjoying the outdoors might begin to displace mere sport hunting. This was particularly relevant in the wake of the decimation of the passenger pigeon in North America—Bailey could remember the sky-darkening flocks of them that he observed in his youth.[28] The new ways of knowing nature that he described in the article sounded very much like nature-study. Photography, in particular, he hoped would begin to supplement hunting as a method for coming into close contact with wild animals in their natural habitats without having to kill them.[29]

The following May issue marked the beginning of the journal's second volume (each volume consisted of six monthly issues), and that issue advertised a new incentive available exclusively to subscribers: a portfolio book of stunning nature photographs by the artists who had been contributors to *Country Life in America*, "featuring text by the editor," titled *Nature Portraits: Studies with Pen and Camera of Our Wild Birds, Animals, Fish and Insects*, and also published by Doubleday. The "studies with pen" would be Bailey's own literary contribution. He took his essay on "The New Hunting," and combined it with a poem, titled "Utility," and several new essays: "The Point of View Towards Nature," "Must a 'Use' Be Found for Everything?" "Science for Science's Sake," "The Extrinsic and Intrinsic Views of Nature," and "The Poetic Interpretation of Nature." The volume was particularly literary as well as philosophical, seeming to channel the Transcendentalist writers of a prior generation and written with a Thoreauvian wit and easy narration—for example, as Bailey describes sitting in on a lesson with a teacher who insists that every adaptation in nature can be explained as having a "use," he writes: "I wondered what would happen if some inquisitive child were to ask what becomes of all the plants that have no thorns or hairs or poison or ill scent. [. . .] As I wondered, a little hand went up. The teacher granted a question. 'Pigweeds ain't got prickers,' said the boy. I saw that the boy was a philosopher" (II.IV). Also like Thoreau, Bailey always grounded his more transcendental observations in the earth of experience, as he would later write in *The Nature-Study Idea*: "The best thing in life is sentiment; and the best sentiment is that which is born of the most accurate knowledge. I like to make this application of Emerson's injunction to 'hitch your wagon to a star'; but it must not be forgotten that a person must have the wagon before he has the star, and he must take due care to stay in the wagon when he rides in space" (I.III).

In the new volume, while "The New Hunting" remains primarily focused on its subject, the other essays all center around children's education, but Bailey insists that his aim in publishing them is to challenge long-ingrained points of view on the natural world. These "points of view are so early impressed upon us," he explains, "that I have purposely chosen

THE pictures in this striking portfolio are a revelation to any one who has not kept up with this nature photography movement. The infinite patience, the arduous labor and the immense technical skill required to get such views of the real live birds, animals, fishes and insects, are evident to all.   Taken from the best work of such men as A. Radclyffe Dugmore, W. E. Carlin, A. G. Wallihan, Herbert K. Job and L. W. Brownell, they are real, not an artist's fancy, and they have the fascination of the actual wild life.   Nothing like them has ever been published in such an attractive form before.   There are fifteen full-page plates: reproductions in full color, photogravures, photographic plates, showing deer, antelope, wild ducks, salmon, Kadiak bear and fox, woodcock on nest, mouse and young, owls, and so on.

12 x 18 inches ; 48 pages, with about 100 half-tone cuts and 15 superb
plates, ready for framing ; all in an attractive portfolio   ·   · $6.00

Figure 2. Advertisement for *Nature Portraits*. *Country Life in America* 2, no. 1 (May 1902): viii. Author's collection.

many illustrations from the teaching of children."[30] Perhaps, but all of these, minus the poem and plus "An Outlook on Winter" and much of the material from his February editorial, were destined for *The Nature-Study Idea* and may well have originated in that first rejected manuscript. In light of Macmillan's rejection, it is possible that Bailey thought this art book would be their final home.

William James Beal and Science versus Sympathy

One of the great ironies in the history of *The Nature-Study Idea*, a book that drew so much inspiration from Bailey's childhood teacher, was that it would take the critical attacks of one of his other favorite teachers to provide him with the impetus to return to it and finally see it through to publication. That teacher was the botanist William James Beal. A precocious teenaged Bailey had originally been drawn to study botany at the college level after successfully inviting Professor Beal to deliver a lecture in South Haven—an invitation motivated in part by the possibility of hosting Beal at the Bailey farm home and using that opportunity to clear up some botanical questions with the eminent scientist.[31] When Bailey eventually did go off to the State Agricultural College, Beal became a favorite professor and an important mentor. Beal, in turn, had been a student of Louis Agassiz, the famous Swiss scientist who took a post at Harvard and would often be credited as one of the primary pedagogical forerunners of the nature-study movement. Agassiz's famous dictum had been, "Teach nature, not books," and in that vein Beal would start a college course in botany with no books, distributing to each student instead a simple twig (likely gathered by a young assistant like Bailey). He would then ask his students to study the bare twig itself for as long as it took until they "saw" something (such as the geometrical pattern of the twig's buds), giving no further instruction.[32] Bailey's exercise with the pussywillow branches that William Donley would so fondly recall was likely based on similar Agassizean lessons learned from Beal. The goal was to train the students' observational skills and develop their capacity for discovery, rather than merely accepting the stated facts in books—such facts often needed revision or challenge, anyway—and to learn directly from a bit of nature.

For no small reason, then, Professor Beal, student of the great Agassiz, considered himself to be an authority on such teaching. Beal understood this "nature-study," as he believed Agassiz had, to be the ideal way to teach *science* and to make students into scientific observers. That, in his mind, was the primary goal of nature-study. He had, moreover, come up at a time in which the sciences were still carving out a space for themselves in the American university, so any movement that could defend the place of science within education seemed critical to Beal.

Yet in the 1890s, the dynamic was shifting. With the dawn of the Progressive Era, scientific advancement became increasingly central to how America thought of itself, and younger scientists were able to pursue their work less defensively than the old guard of Beal's generation had. As Cornell's nature-study movement coalesced, its leaders were less concerned about defending science than they were in defending the value of rural life

Figure 3. The staff of *The College Speculum*, undated. Bailey served as founding editor of this student-run paper published out of the Michigan Agricultural College, seen here sitting directly beside his professor, William James Beal, far left, who served as science editor for the paper. The first issue, in August of 1881, included a short description by Beal of his time as a student of Agassiz and the hands-on instruction Agassiz gave his students. This photo would have been taken in Bailey's senior year of 1881–1882. Courtesy of the Michigan State University Archives and Historical Collections.

and communities in the face of rapid urbanization—or even, for Bailey, defending the broader idea that life anywhere became worthwhile once one learned to interpret and understand the natural world in which one lived. In 1897, in the first full year of the Cornell Nature-Study Bureau's publishing effort, Bailey had published a little leaflet under the title "What Is Nature-Study?" to attempt to bring clarity to the fledgling movement and articulate the aims that he had observed among its successful practitioners—that leaflet is reproduced in this volume. While nature-study clearly contributed to scientific learning in the classroom among young students, Bailey insisted that the movement's aim was not narrowly scientific:

> It is the seeing of things which one looks at, and the drawing of proper conclusions from what one sees. Nature-study is not the study of a science, as

of botany, entomology, geology, and the like. That is, it takes the things at hand, and endeavors to understand them, without reference to the systematic order or relationship of the objects. It is wholly informal and unsystematic, the same as the objects are which one sees. It is entirely divorced from definitions, or from explanations in books. It is therefore supremely natural. It simply trains the eye and the mind to see and to comprehend the common things of life; and the result is not directly the acquirement of science but the establishment of a living sympathy with everything that is. (Related Writings, What Is Nature-Study?)

That word—"sympathy"—became a lodestar for the Baileyan arm of the movement. Through training the powers of observation ("seeing," as he writes here, channeling Ms. Field's early lesson—but he would also come to emphasize the involvement of all five senses, as he describes in I.III, this volume), Bailey believed that nature-study uniquely brought the student into sympathetic relation with the observable world, which is the world of nature, whether cultivated or wild. The word *sympathy* at that time was still understood in its fuller sense, something like what today is more often described as *empathy*, a *full resonance* with something outside oneself. (Think, for instance, of the "sympathetic" strings on a sitar, which produce their own sound by vibrating along with the sound waves produced by the plucked strings above them.)

To Beal, such an emphasis on sympathy was a danger to science. He joined a growing chorus of academic scientists who were becoming anxious about the nature-study publications circulating among teachers, and he critiqued their "sentimentality." Most distressing was that writers like Bailey emphasized that any motivated teacher with a little bit of knowledge of the natural world could teach nature-study—that they need not be trained scientists themselves to open the eyes of their pupils. For Bailey, this was liberating for the student as well as for the teacher:

> You simply go as far as you know, and then say to the pupil that you cannot answer the questions which you cannot. This at once elevates you in the pupil's estimation, for the pupil is convinced of your truthfulness, and is made to feel—but how seldom is the sensation!—that knowledge is not the peculiar property of one person, but is the right of any one who seeks it. It ought to set the pupil inquiring for himself. (III)

Beal worried that teachers who lacked a college degree in the sciences, motivated by a desire to cultivate sympathy among their students with the natural world, would "sentimentalize" the natural world and teach a watered-down or romanticized version of natural history.

Behind much of this critique lurked a latent sexist assumption that a movement arising from public school teachers, most of whom at the time were women, would be sentimental and unscientific. (Teaching was one of the few professions in which women could respectably work to support themselves at that time, often just for a few years before marriage, and it was low-paying compared with many jobs available to men. The "blessed year" that Bailey remembered with Julia Field may well have been her only year of classroom teaching.) Bailey insisted that the movement did originate with these teachers, but he firmly denied that because of that the instruction would devolve into sentimentality. It was no small coincidence that he hired Anna Botsford Comstock from the very beginning as he was building the nature-study program, making her the first woman professor at Cornell (although she was denied the full faculty salary, which Bailey had requested for her, by the Board of Regents for years).[33] Shortly afterward, at Comstock's encouragement, Bailey went on to hire the pioneering Martha Van Rensselaer to chair the new Department of Home Economics, and the nature-study program faculty was soon dominated by visionary and talented women, including Alice McCloskey, Ada Georgia, and the sisters Julia Rogers and Mary Rogers Miller.[34] Comstock would later describe Bailey's leadership of this team as "sympathetic and understanding."[35] This extended to his students as well as to his colleagues; as he reassured one young woman who was an advanced student of his, "You quite underestimate your own knowledge because you feel the responsibility of your work."[36] Bailey was of a younger generation than Beal, and he could see through some of his old mentor's blind spots to recognize that these women were not watering down science; they were opening the eyes of the young, in an increasingly rigorous and well-developed pedagogical method, to learn from the world around them through experience and self-discovery. They were creating lifelong learners, like Julia Field had done with a young Bailey so many years before, and they were doing so among the many in the public schools, not just among the few who would go on to study science in college. This was a movement of the common schools and of the common people, and these women were its leaders.

Figure 4. The faculty and students of the Nature-Study Summer School at Cornell, circa 1898. The nature-study program's first director, Isaac P. Roberts, is third from left, front row, hat in hand. Right of Roberts, between two children, is "Uncle" John Spencer. Just behind Spencer and to the right is John Henry Comstock, to the right of him is Bailey, and above Bailey and to the right is Anna Botsford Comstock. Among the teachers here to take the course, the number of women is notable. Courtesy of the Division of Rare and Manuscript Collections, Cornell University Library.

Beal's response came in the "correspondence" section of the journal *Science* in 1902, just a year after Bailey's manuscript had been rejected. In two installments, appearing in June and December, Beal appropriated Bailey's title, "What Is Nature-Study?," as the heading of his two-part rebuttal. (The rebuttals are also reproduced in Related Writings, this volume.) The first installment quotes Bailey's original essay of the same name and contrasts the quotation with one from Dietrick Lange, the supervisor of nature-study for schools in St. Paul, Minnesota, who argues that "[n]ature study [. . .] is understood to be the work in elementary science taught below the high school" and warns that "[g]ushing sentimentalism or mere rambling talks will be as barren in results as undigested statistics." Beal proceeds to describe his own credentials and training under

Agassiz, noting that "[t]hrough these students of Agassiz and their students down to the third generation"—meaning Bailey's generation—"this spirit of independent work has come filtering along for fifty years or more." However: "With much that is good in nature study comes much that is positively injurious, and unfortunately large numbers are unable to distinguish between the true and the false." While Beal refrains from directly attacking his former student on this point, the conflation of Bailey's "wholly informal and unsystematic" approach with what Lange describes as "mere rambling talks" appears to be quite clear, as are the dangers to Bailey's audience of teachers who Beal implies may be "unable to distinguish between the true and the false."

He concludes with an image, meant to disturb the readers of *Science*, of a young nature-study student who produced two drawings of bees visiting apple blossoms, which are all wrong: "The bees are not alike; each has two wings only; the heads and legs are unlike anything ever attached to bees. The apple blossoms are five-lobed (gamopetalous), with three stamens growing from the base of each lobe of the corolla." Such drawings, Beal concludes, "are absolutely worthless, in fact injurious, to any young person who makes them or even looks at them."[37]

"It Is Spirit": The Book Finds Its Moment

Bailey was no stranger to the pages of *Science* and had written for the journal's correspondence section before. He could have sent in his own rebuttal there, defending his idea of nature-study against these accusations of "gushing sentimentalism." If he had, perhaps the scientific community would have seen the spat as another unresolved quarrel within the ranks. But Bailey knew that the center of the movement was not to be found in the pages of *Science* or among its readers. His work had brought him into close contact with teachers in the public schools, and he knew how to reach them. He knew the pulse of the movement, and he knew that he didn't need to mount any defense. He was prepared to launch a manifesto.

Beal published the first of his two critical essays in *Science* just a month after *Nature Portraits* was first advertised in the pages of *Country Life in America* in May. Here was a challenge that the promotional folio of essays and photographs would not suffice to answer. The final form of

PROF. L. H. BAILEY'S NEW BOOK
THE NATURE-STUDY
IDEA
By LIBERTY H. BAILEY

Professor Bailey speaks with peculiar authority on this subject from the vantage ground of his educational work, his great horticultural cyclopedia and other outdoor books, and his position as editor of COUNTRY LIFE IN AMERICA. This volume is an illuminating and suggestive study of the whole movement toward Nature which has been so marked a characteristic of the last ten years. (About 150 pages, cloth binding. Price, net, $1.00.)

Other Beautiful Nature Books
WITH COLORED PLATES AND PHOTOGRAPHS FROM LIFE

AMERICAN ANIMALS
By WITMER STONE & WM. EVERITT CRAM
6 color plates and more than 100 photographs from life. ($3.00 net.)

AMERICAN FOOD AND GAME FISHES
By DAVID STARR JORDAN & BARTON W. EVERMANN
10 color plates, 100 photographs of live fish in the water, and 200 text cuts. ($4.00 net.)

THE INSECT BOOK
By DR. LELAND O. HOWARD
16 colored plates, 32 black and white. ($3.00 net.)

THE BUTTERFLY BOOK
By DR. W. J. HOLLAND
48 colored plates. (Special net, $3.00.)

THE MUSHROOM BOOK
By NINA L. MARSHALL
24 colored plates, 24 black and white, and about 100 text cuts. ($3.00 net.)

DOUBLEDAY, PAGE & COMPANY, 34 Union Sq., New York

Figure 5. Advertisement from *Country Life in America* 3, no. 5 (March 1903): cxlix. Author's collection.

The Nature-Study Idea probably began to take shape during that summer or fall of 1902. The essays from *Nature Portraits*, together with his "Outlook on Winter" essay from *Country Life in America*, came together with light revisions to form the poetic and narrative central section of the book, Part II—the section that best *shows* the reader what nature-study looks like in all of its aesthetic, spiritual, and personal dimensions. Part I drew on notes from the many lectures he had delivered to teachers' groups and summer courses on nature-study, comprising a series of essays that more clearly set forth the guiding philosophy and ideals of the nature-study movement—that section most clearly *tells* the reader what nature-study is, how it first developed, and how it is practiced.[38] He then concluded the volume with a practicable but lively series of "inquiries and answers" that had commonly arisen in his talks with teachers, allowing him to fill in additional gaps and create a resource that would be even easier for working teachers to consult. How much of this marked a change from the original 1901 manuscript and a response to Beal's critique cannot be determined, but the book came to represent the product of over six years' writing and work in the field that was simply waiting to be brought together. Beal's second installment, in December, assembled seven testimonials from "eminent scientific men" (and one woman) that all seemed to undermine the work of the Cornell nature-study team; this may have been the last straw. With the publication of Beal's critiques of the movement,

Bailey not only had his manifesto on nature-study, he had a real impetus for publishing it, too.

While he does not refer to his former professor by name anywhere in the book, Bailey does position it as a direct response to Beal's criticism on the very first page. For the opening chapter, he reclaims the title of his 1897 leaflet (which Beal had appropriated for his *Science* articles): "What Is Nature-Study?" This book, the reclaimed chapter title implies, would finally provide the definitive answer. He then proceeds to address the second of Beal's two critical articles—dismissively—in the book's first paragraph: "A contributor to a recent issue of a leading technical journal has endeavored to find a satisfactory answer to the question, 'What is nature-study?' by appealing to 'eminent scientific men.' The answers of these men are printed there in full." The description of *Science* as "a leading technical journal," and of one of the country's foremost botanists as simply "a contributor" to it, represent a rather shocking dismissal from Beal's star student and a radical alignment of thought with working schoolteachers rather than with the academic intelligentsia. He then offers one of the primary thesis statements of the book: "Now, the nature-study movement is not a product of 'eminent scientific men,' nor directly of the current natural-science movement. It is a product of the common schools." By the next paragraph, the thesis has expanded: "[Nature-study] designates the movement originating in the common schools to open the pupil's mind by direct observation to a knowledge and love of the common things and experiences in the child's life and environment. It is a pedagogical term, not a scientific term." And finally: "Nature-study, then, is not science. It is not knowledge. It is not facts. It is spirit. It is an attitude of mind. It concerns itself with the child's outlook on the world."

With that opening salvo, Bailey is off to the races, with all the confidence of a writer who writes from experience and knows that he knows his subject better than his harshest critics. The book was published in April of 1903, just four months after Beal's second essay had appeared in *Science*, and, while it would remain controversial, it quickly became a standard in the field, with a particularly enthusiastic audience among teachers. It would also become more influential than almost any other book in shaping the nature-study movement over the ensuing decades.

A FINE SET FOR NATURE LOVERS
The Little Nature Library

WE are anxious that ten thousand more people shall take COUNTRY LIFE IN AMERICA before the snow flies. To induce readers to become regular subscribers, we have arranged for a new set of books which appeal particularly to those who are interested in the country, and we will send them on approval so that there may be no chance of dissatisfaction. Here is the proposition:

We will send "Country Life in America" to you (or to some friend if you are already a subscriber) for a full year, post-paid, and this set of six volumes in an attractive box:

THE NATURE-STUDY IDEA   By Liberty H. Bailey
A suggestive analysis of the movement which has grown so stupendously in the last few years. It points the way for the individual, and will be welcomed from such an authority as Professor Bailey, who stands foremost as a leader of Nature-Study.

HOW TO ATTRACT THE BIRDS   By Neltje Blanchan
Author of "Bird Neighbors," "Birds That Hunt and Are Hunted," "Nature's Garden," etc., etc.
These intimate, suggestive and charmingly written chapters are ornamented with a great number of extraordinary photographs, and form an altogether unique work on the almost untouched subject of "making friends" with the "bird neighbors" to whom the author has introduced so many thousands of readers.

NATURE AND THE CAMERA   By A. Radclyffe Dugmore
From the choice of a camera to questions of lighting and to the problem of "snapping" shy birds and animals in their native haunts—every step is explained so simply as to be easily comprehended, even by the beginner.

THE BROOK BOOK   By Mary Rogers Miller
"Lovingly intimate with its mysteries, very near to the heart of the brook, are these pages. They cannot help but charm all who have the love of nature in their hearts."—*Pittsburg Gazette.*

AMONG THE WATER-FOWL   By Herbert K. Job
"This book ought to make thousands of sportsmen throw away their guns and follow the birds with an implement which requires more eyes, brains and heart to make a successful shot. The pictures are as good as the text, and the latter ranks with the best of its kind."—*The Philadelphia Era.*

NATURE BIOGRAPHIES   By Clarence Moores Weed
A sort of personal acquaintance with the lives of the more common butterflies, moths, grasshoppers, flies, etc., told in a most entertaining way.

THE ILLUSTRATIONS
of which there are about 500, are mostly photographs from life and are the most beautiful that have ever been published. They are remarkable as intimate pictures of natural life.

For $100 down and $1.00 a month for seven months or $7.20 cash if you desire to pay at one time.

Doubleday, Page & Company, Publishers,
34 Union Square, New York

Figure 6. Advertisement for The Little Nature Library. *Country Life in America* 4, no. 2 (June 1903): 99. Author's collection. The promise of lavish illustrations would have been misleading only for Bailey's contribution, which contained only two illustrations, drawn by children.

"For All Things Are of Kin" and "Man Is a Land Animal": The "Earth-Philosophy" of Nature-Study

From the beginning, it seems *The Nature-Study Idea* always aimed for a large audience and a broad impact. The problem of cultivating a new generation that is more sympathetic to the more-than-human world around them, and that will better care for that world, turns out to be one that concerns everyone. While the text's three-part structure and pragmatic organization reflect the need to be easily and repeatedly consulted by working teachers, the prose and subject matter are also engaging and accessible enough to delight and inspire the general reader of nature books. The peculiarly personable tone that runs through most of the work, sometimes whimsical and other times lofty and spiritual, speaks to a concern that far exceeds the classroom and reaches to more fundamental questions about the world that adults will hand to their children and the ways in which we might best make sense of that world. In their marketing efforts, Doubleday initially packaged the book in their Little Nature Library series (see figures 5–7), with Bailey's book providing a philosophical lens to help readers understand how to use the more subject-specific (though also companionable and literary) books, like *The Brook Book* by Bailey's nature-study colleague at Cornell, Mary Rogers Miller, or *Nature and the Camera*, by the photographer A. Radclyffe Dugmore, whose photographs had been prominently featured in *Nature Portraits* and throughout the issues of *Country Life in America*. What Bailey sought to articulate was a broader outlook on the more-than-human world, born of his experiences growing up and living in a working relationship with the land on farms and in rural spaces, that seemed to him, at the dawn of the twentieth century and in the midst of American industrialization, more important than ever for children and adults alike.

The success of *The Nature-Study Idea* finally opened the door for Bailey to publish what would become a long series of books that we might today describe as environmental philosophy, although he perhaps came up with the better name, "earth-philosophy."[39] He was concerned about the effects of industrialization on people's lives, and he could draw the line that linked lack of understanding of the natural world to the destruction of that world. In 1915, he would make that line clear in *The Holy Earth*, describing a "habit of destruction" born of a misunderstanding of humanity's fundamental relationship with the rest of the world—a relationship defined both by evolutionary kinship with the full "bio-centric" creation, as he called it, but also by the responsibility implied by humanity's outsized influence on the ecosphere, an influence he described as one of inescapable "dominion" (borrowing language from Genesis), demanding of

Figure 7. Advertisement for The Little Nature Library. *Country Life in America* 4, no. 5 (September 1903): 310. Author's collection.

us an ethic of humble stewardship.[40] While Bailey had watched the extermination of the passenger pigeon, he had also grown up among the members of the Pokagon Band of Potawatomi, and as a child he was taught by Potawatomi children how to catch passenger pigeons in ways that their ancestors had been sustainably practicing for generations. "I knew the Indians," he would later say, "and I picked up something of their outlooks."[41] That influence may have been larger than he ever truly let on, judging by the philosophical passages in his early *Onamanni* manuscript. In *The Holy Earth*, he would attempt to frame that ethic of caring stewardship in the terms of the Judeo-Christian tradition and the agrarian side of his upbringing, finding there, too, ample resources for a more sustainable outlook to the human place in and responsibility toward nature. Bailey believed that Western culture had not yet realized that ethic in practice, and he called on his readers to "take a new hold" on their relationship with the rest of the world.[42]

The most direct means to begin bringing about this new hold, Bailey believed, would be through reform at the level of education—no quantity of environmental safeguards would last if the upcoming generations lacked a knowledge of and genuine appreciation for the natural world. In this way, nature-study was foundational to the larger vision of his earth-philosophy. Part of the new hold was a radical acceptance of the revelations of science, and for Bailey the science of evolution was particularly powerful in communicating the *familial* human relationship with and responsibility to all life on Earth. While he continued to maintain that nature-study was not itself science—that it embraced a whole spectrum of methods to relate sympathetically to the natural world, from geology to poetry, but ideally all rooted in firsthand observation—he also argued that the Cornell approach to nature-study, emphasizing sympathy and love of nature, was a distinct *benefit* to the sciences, not the danger that Beal saw it as.

For instance, in one of the book's most delightful chapters, "The Integument-Man," Bailey relates the story of a scientist who had read his influential nature-study leaflet titled "How a Squash Plant Gets Out of the Seed" (reproduced in Related Writings, this volume) and later complained to him that the leaflet misled the public in claiming that a squash plant gets out of its *seed* rather than out of its *integument*. The quibble echoes Beal's concern about the improperly gamopetalous flowers drawn by the young child and the threat that the picture might make "to any young person who makes them or even looks at them." Bailey has a bit of fun with the anxiety: "The Integument-Man is afraid that this popular nature-study will undermine and discourage the teaching of science," he writes, as if he could be referring to Beal personally. "Needless to say, the fear is absolutely groundless. [. . .] Science-teaching has more to fear from desiccated science-teaching than it has from nature-study. It is the Integument-Man himself who is discouraging the teaching of science." The worst threat that the Integument-Man poses, according to Bailey, is that he discourages teachers from taking their children outdoors to learn from nature "because of a lack of technical knowledge of the subject." Such hesitation might come from a good place—as in the case of the education student, mentioned above, whom Bailey had to reassure, stating that she underestimated her own knowledge only "because you feel the responsibility of your work"—but it misses the point, because "technical knowledge of the subject does not make a good teacher" (I.IV).

Moreover, the science would serve the sympathy, rather than the other way around—"science for science's sake" had little place in the elementary classroom, in Bailey's mind—and it was the sympathy that would form the bedrock on which the nascent country life movement could build a more sustainable world. The evolution of his chapter "Nature-Study Agriculture," which Bailey almost completely rewrote (the original may be found under Major Sections Restored, this volume), shows the maturation of his thought along these lines. That the Cornell program would root its work in rural schools made a certain sense even beyond the university's land-grant mission: "Man is a land animal," Bailey writes, "and his connection with the earth, the soil, the plants, animals and atmosphere is intimate and fundamental. This earth-relationship is best expressed in agriculture,—not agriculture merely as a livelihood, but as the expression of the essential relationship of man to his planet home" (I.VII). Agricultural topics would be instructive to any student—food and clothing, after all, are as much a part of the urban neighborhood as the rural one. But nature-study seemed an especially important foundation for future farmers: "The best agriculture is a perfect adaptation of man to his natural environment. [. . .] There is no effective living in the open country unless the mind is sensitive to the objects and phenomena of the open country; and no thoroughly good farming is possible without this same knowledge and outlook. Good farmers are good naturalists" (I.VII). And good farmers would be measured by their impact on the land and community, not by their economic efficiency alone. As he would write just a few years later, "[T]he requirements of a good farmer are at least four: The ability to make a full and comfortable living from the land; to rear a family carefully and well; to be of good service to the community; to leave the farm more productive than it was when he took it."[43]

The benefits to land, community, and child that would come from the nature-study movement as Bailey envisioned it also crossed racial lines, which was particularly significant in light of the fact that many proponents of the more narrowly scientific approach to nature-study (including many of the "eminent scientific men" Beal had cited) were also prominent eugenicists. Eugenics—the long-debunked but still pernicious theory that human racial characteristics reflect an evolutionary hierarchy, in which people considered white sit at the top—depended both on a linear understanding of evolution, in which each rung up the ladder put distance between "higher" and "lower" forms of life, and on a confusion of cultural and biological forces and characteristics. Bailey seems to have been skeptical of this simplistic understanding of the evolution of life, as he illustrates in his chapter "Must a 'Use' Be Found for Everything?" If any lessons should be taken into social life from the study of biological evolution, Bailey indicated that they should stress the revelation of our incredible kinship and familiarity with all life rather than relatively arbitrary differences. Regarding nature, he writes that "[i]t were better that we know the things, small and great, which make up this environment, and that we live with them in harmony, for all things are of kin; then shall we love and be content" (II.I). Carried into the human realm, this emphasis on kinship would seem to contradict the racist anxieties about difference that lay at the center of eugenics. In terms of his personal life, we know that Bailey mentored African American students, with whom he sometimes maintained a close relationship for many years.[44] The very year *The Nature-Study Idea* first appeared, Roscoe Conkling Bruce of the Tuskegee Institute wrote to Bailey asking him "if among your recent graduates there is a Negro capable of teaching elementary Agriculture," and within a week he wrote back thanking Bailey for the speedy response and the two strong recommendations.[45]

And the nature-study movement had already taken hold among the segregated educational institutions serving Black students in the South— just a year after *The Nature-Study Idea* appeared, Booker T. Washington's book, *Working with the Hands*, would articulate a remarkably similar philosophy that had already been applied at the Tuskegee Institute's practice school. Washington's sense of the movement seems perfectly aligned: "I believe that the time is not far distant when every school in the rural districts and in the small towns will be surrounded by a garden, and that one of the objects of the course of study will be to teach the child something about real country life, and about country occupations."[46] The arguments for nature-study, including the deepening of the child's happiness through empowering the child to pursue an intellectually stimulating life on the land, found a ready reception among both Washington's and Bailey's readerships. While it is not clear how much these two heads of agricultural colleges corresponded, Washington would contribute an essay on "The Negro Farmer" to Bailey's *Cyclopedia of American Agriculture* in 1909.[47] The famous agriculturist George Washington Carver would eventually lead the nature-study program at Tuskegee, and he would also sit on the editorial board of *The Nature-Study Review* at Bailey's request, in which role he "encouraged creativity and aesthetic appreciation as well as economic innovation."[48]

In the text of *The Nature-Study Idea*, Bailey would specifically cite the nature-study work underway at the Hampton Institute, another southern institute of higher learning that served Black students. He had published a short article by Jean E. Davis, an instructor at Hampton's Whittier Training School, in the March 1903 issue of *Country Life in America*, and in *The Nature-Study Idea* he reprinted it in Part III to respond to a question about how to organize an effective school garden. While Davis's article is tainted by racializing language that should stand out to us today, the inclusion of her work at the Whittier School indicates that Bailey and others saw nature-study as an empowering and uplifting movement across racial lines. The work of Black nature-study advocates like Washington and Carver forms a major part of the legacy of more recent efforts in the struggle for land and food sovereignty, as well as access to quality outdoor education, among communities of color.[49]

Figure 8. Frances Benjamin Johnston, Whittier School students on a field trip studying plants, Hampton, Virginia, 1899. Courtesy of the Library of Congress Prints and Photographs Division, Washington, D.C.

Ever balancing pragmatism with optimism, Bailey believed that if teachers and parents could just be encouraged to lead their students and children toward direct and self-guided learning in the outdoors, the outlook to nature would evolve and people would move toward "the new hold" over the course of the unfolding generations. The practical task of the book, then, would be to offer that inspiration and empowerment, and that inspirational mission is likely a major reason why *The Nature-Study Idea* was so popular among teachers in the early twentieth century. It reassured those who had a gift for teaching that their instincts were to be trusted and not limited by the suspicions of an increasingly standardized educational landscape and an industrializing social mind-set—suspicions that have become only more pronounced today—and it did so in ways that continue to ring true. "Then teach!" Bailey exclaims at the end of "The Integument-Man":

If you love nature and have living and accurate knowledge of some small part of it, teach! Do not fear your scientific reputation if you feel the call to teach. Your reputation is not to be made as a geologist or zoölogist or botanist, but as a leader. When beginning to teach birds, think more of the pupils than of ornithology. The pupil's mind and sympathies are to be expanded: the science of ornithology is not to be extended; the science will take care of itself. Remember that spirit is more important than information. The teacher who thinks first of his subject teaches science; he who thinks first of his pupil teaches nature-study. With your whole heart, teach!

Do not be afraid of the Integument-Man. (I.IV)

It is no surprise that Bailey would become what Anna Botsford Comstock called "the inspiring leader of the [nature-study] movement," nor that his new manifesto would soon be more widely known and cited than his earlier textbooks. In fact, in 1906, *The Nature-Study Review* listed Bailey alongside Clifton Hodge as the "top-ranked authors" in a list of the ten best nature-study books.[50] (It would be five more years before Comstock would publish her *Handbook of Nature-Study*, which would then become the bestseller in the field.) Sally Gregory Kohlstedt, in her definitive history of the American nature-study movement, notes that Bailey became "perhaps the most publicly recognized name in nature study," and that *The Nature-Study Idea*, "representing the philosophy of the subject rather than practical guidelines for its implementation" and serving "as a kind of inspirational text," was "widely read by teachers and often assigned in normal school courses."[51] Teachers, professors of education, and other readers were drawn to the book because the philosophy it espoused resonated with them and worked in practice, but also because it called them to "take a new hold" on how they viewed their own place in the natural world and what that meant for how children should be raised and taught.

These provocations remain as relevant as they did over a century ago when they first appeared. In many ways, they feel even more urgent. While Bailey could be painfully aware of the costs of species extermination and soil degradation, he could not have imagined the havoc that the "habit of destructiveness" would bring to the very fabric of the ecosphere as atmospheric CO_2 levels continued their climb over the course of the ensuing century. He would also doubtlessly find it painful to learn the ways in which the American educational system has turned further toward information-cramming and high-stakes testing, particularly in the elementary grades, rather than focusing on establishing fundamentals and providing ample time and space for self-directed exploration and discovery. But today, as in Bailey's time, there are still many teachers, parents, and others who push against these trends and seek to provide every opportunity for children to learn from firsthand contact with the world in which they live and who see the continuity linking such efforts with the conservation of the earth and the growth of a well-educated democracy. For those leaders in their efforts, a return to Bailey's vision may still prove revolutionary.

"The Spirit Will Live": A Book Revised, the Idea Continued

As his literary efforts with Macmillan matured, Bailey began publishing more and more works of "earth-philosophy" with them, and in 1908 they indicated a desire to purchase the publishing rights from Doubleday to the book they had rejected just seven years earlier. Evidently Doubleday was reticent to sell the rights to the volume, leading George P. Brett at Macmillan to suggest writing an entirely new book for them. Bailey's response illuminates both the value he had come to see in his book and also his plan for it:

> The only satisfactory way to handle 'The Nature Study [*sic*] Idea' is to secure all rights from Doubleday, Page & Company. This has now become a standard pedagogical work and is much quoted. I do not care to make any change in the phraseology in a number of the most important parts. I wish to eliminate some of the parts which are now somewhat out of date, and to add some new ones. To write a new book would not at all take the place of the old one.[52]

Bailey did take the opportunity to give the book a major overhaul, and it joined three of his other philosophical books, *The Outlook to Nature*, *The State and the Farmer*, and *The Country-Life Movement in the United States*, to form what he called "The Rural Outlook Set." The revised Macmillan edition of the book appeared in 1909, and, matching the others in the series, it was "made a 16 Mo," just like Bailey had asked in 1901 (although lacking the "untrimmed edges" he envisioned). It may even have met his hopes of being "put up [. . .] in an artistic way," to the extent that it was gilded along the top of its pages and featured typographical flourishes more characteristic of a literary work than the textbooks Macmillan had long published for him.

The revision of *The Nature-Study Idea* and the appearance of The Rural Outlook Set also corresponded with a moment in which Bailey was seriously taking stock of his career and the possibility of breaking away from the deanship of the New York State College of Agriculture at Cornell to strike out as a full-time author and editor with Macmillan. As early as December 1907, he wrote to Brett, "If I decide to sever my connection here, it should be possible for me to catch up all our delayed plans and to begin a large constructive program. The determining factor in my decision will be the prospect of the books and contracts I have with you; and this is the best testimony I can give you of my faith in The Macmillan Co. The relations with you have always been most satisfactory."[53] By April 1909, Bailey and Brett had had a personal meeting in New York City to discuss a possible arrangement for Bailey to throw himself full-time into his publishing work and step away from his position at the college, and in a flurry of correspondence over the ensuing months the two sought to negotiate a contract at the same time that Bailey worked to complete the *Nature-Study* revisions. Having just served as chair of Roosevelt's Commission on Country Life and pushed through a remarkable number of edited and authored volumes in the preceding year, Bailey was at the height of his powers and also feeling increasingly strained by a workload divided between the college, the nation, and the publishing house. When he arrived in New York to meet with Brett, he came directly from a meeting with the newly inaugurated President William Howard Taft in Washington, D.C., presumably over the future of the commission's work (which was destined to be scuttled in Taft's term—by the time of Woodrow Wilson's administration, its archive of over one hundred thousand letters and completed circulars from rural Americans would be burned on the orders of Wilson's secretary of agriculture, D. F. Houston).[54] Things escalated quickly; in a dramatic letter on May 14, Bailey began by stating that he was formally terminating his arrangement with Doubleday to produce the revision of *The Nature-Study Idea* for Macmillan "not later than August 1st, so that it may be published by October 1st, as you desire," and then he concluded the letter with this one-sentence paragraph before signing off: "You will be interested to know that I have today placed my resignation in the hands of the President of the University."[55]

Ultimately, it seems, the university was able to retain Bailey for several more years, and he continued in his position as dean until 1913. Many of the books and series he projected to Brett still came to pass: they centered around an expansion of the Rural Science Series, the establishment of the Rural Text-Book Series, and then the establishment of a new series, to be authored by Bailey himself, the future of which may have suffered more than that of the others from his decision to postpone retirement. In May 1909, the shape of this series emerged in his correspondence with Brett as he described his intent to shorten *The Outlook to Nature* in a revision to match the style of *The Nature-Study Idea* and *The State and the Farmer* and indicating that "[f]rom time to time I shall have other personal books of this nature, which will constitute a *Rural Outlook Series*."[56] Several days later, another letter to Brett opens, "You will be shocked to learn that I sometimes write in verse. This habit has been mostly innocuous, but I shall now publish a volume of verse. Of course I prefer to have you handle it. [. . .] I shall call it 'Wind and Weather,' and should like it to be one of the volumes of the *Rural Outlook* kind."[57] Whether Brett ultimately turned down the poetry collection or Bailey simply became too busy in his return to administrative work is not clear, but *Wind and Weather* would have to wait another seven years before finding a home with Charles Scribner's Sons in what appears to have been a second attempt at an ongoing "Rural Outlook Series," titled The Background Books: The Philosophy of the Holy Earth. In the meantime, his proposed series was cut short at four and became the Rural Outlook "Set," with the publication of *The Country-Life Movement in the United States* in 1911.

The Nature-Study Idea continued its work, however, and Bailey continued to be an active spokesperson for the nature-study movement. The nature-study program at Cornell continued to prepare teachers to foster experiential outdoor learning opportunities for their students, and the model it set would be imitated in states and provinces across the United States and Canada—the spirit of the old program is alive in the courses on "nature education" offered today by Cornell's Civic Ecology Lab.[58] The success of their voluminous leaflet series led to the publication by the state of New York of the bound, 607-page anthology *Cornell Nature-Study*

Leaflets in 1904, the year after *The Nature-Study Idea* appeared, which Edward F. Bigelow, editor of *Nature and Science*, declared in a letter to Bailey "literally FILLS A LONG-FELT WANT."[59] The process of collecting and revising together an anthology of nature-study leaflets provided the model for Anna Botsford Comstock when she produced her *Handbook of Nature-Study* in 1911, which she dedicated to her colleagues Bailey and Spencer. When it appeared, Comstock's book met with much greater success than the earlier state-published volume had, and it would go through numerous revisions. While it would be impossible to measure the impact of a book like Comstock's *Handbook*, we do know that it was a foundational text for a young Rachel Carson, whose mother Maria used it to guide their outdoor activities throughout Carson's childhood.[60] It remains not only in print, but one of the best-selling titles in the catalog of Cornell University Press to this day. And Bailey's own textbook-writing days were not over; his *Botany* for elementary schools would in 1913 be lightly revised and retitled *Botany for Secondary Schools*, apparently having been a bit too advanced for elementary use; it was replaced for that purpose by his 1908 *Beginners' Botany*, which in the ensuing decades would also be published by Macmillan in Canadian editions, some of which were tailored by regional editors to the flora and fauna of particular provinces.

The many children's organizations that spun off from the Cornell work also had long afterlives. Under John Spencer's leadership, the Junior Naturalist Clubs reached a total membership of around thirty thousand. These later became known as Junior Citizenship Clubs, and similar rural clubs became increasingly common throughout the country, often known as "boys and girls clubs," coalescing under the national 4-H Club banner by the 1920s.[61] In 1908, Bailey became the founding president of the American Nature-Study Society, a position in which he served a second term after his retirement, from 1915–1917.[62] That organization became the center of the nature-study movement for many years, publishing *The Nature-Study Review* for nearly two decades and continuing to advocate for children's outdoor learning long after Bailey's time—it was distributing newsletters as late as 1999 and currently claims the status of "America's oldest organization for [the] environment," though the current status of the organization is unclear.[63] As many other nature-study organizations dissolved or lost their momentum after the first few decades, Kohlstedt notes that "[t]he stalwarts of nature study proved to be Anna Comstock, Liberty Hyde Bailey, and many of the students and colleagues they had inspired."[64] Bailey's dedication to education by and for women and girls continued as well, and late in his life he donated a portion of his farm outside of Ithaca, known as Bailiwick, to become part of a Girl Scout camp founded there in the 1920s by his neighbor and colleague Comstock.[65] His fieldstone house still stands there as part of the property, which continues to serve the Girl Scouts in the Finger Lakes region of New York under the name Comstock Adventure Center.

Remembering the "Teacher Who Allowed a Boy to Grow"

With all of the success of *The Nature-Study Idea* and the many publications and programs he had been involved with, Bailey never lost touch with the fact that the nature-study movement originated with teachers and that teachers continued to drive the movement's innovation. He also never forgot where his own educational inspiration came from, as is evident in the revised edition's dedication to "Mrs. Julia Field-King, a teacher who allowed a boy to grow." In fact, he was so serious about his indebtedness to his childhood teacher that he managed to locate her in 1909 where she had resettled in the town of Cheltenham, England, apparently either widowed or separated and living with a friend in a house which still stands at Longleat on Queen's Road, and he wrote to her there.[66]

Julia Field-King's voice would have been lost to history if not for Bailey's dedication in *The Nature-Study Idea* and his descriptions of her in published speeches. Her archival record, however, paints a portrait of an accomplished woman driven by ambition and independent thinking. Julia A. Field was born in Chicago, around the year 1840. She attended Elmira Collegiate Seminary, considered to be the first women's college in the United States that offered degrees comparable to those available to men, and Oberlin College, considered the oldest coeducational liberal arts college in the country.[67] It is not clear when or why she moved to the village of South Haven, just across Lake Michigan from her hometown and where young Lib Bailey would be one of her pupils at the two-room Central School, but later, at the age of thirty-eight, Julia Field married Hervey King on September 12, 1878, back in her hometown of Chicago.[68] That would have been the fall of Bailey's sophomore year at the State Agricultural College in Michigan.

Marriage would have marked the end of many a young woman's brief working career in the nineteenth century, but it may have been during this period that Field-King enrolled at the Hahnemann Medical College in Chicago, a school that taught homeopathic as well as traditional medicine and became coeducational in 1871.[69] At any rate, an evolving interest in alternative medicine may have been what led her in 1888 to apply to the Massachusetts Metaphysical College in Boston—the school founded by Mary Baker Eddy to train the leaders of her new religious sect of Christian Science, known for the practice of "spiritual healing." Field-King described herself on her application form as a "Homeopathic Physician," trained at Hahnemann.[70] While homeopathy was considered an alternative form of medicine in the late nineteenth century, it was also respected, and homeopathic physicians were also trained in traditional medicine. Field-King's late life appears to have been dominated by Christian Science, and from her voluminous correspondence with Eddy, it seems she was committed to this sect that, in her view, embraced both empirical science and individual spirituality over dogma.[71] It was also an idealistic, increasingly influential organization founded and led by a visionary woman whom Field-King clearly admired. Eventually, Field-King gained the title C.S.D., or Doctor of Christian Science, the highest degree awarded by the organization. She also began to write and publish pamphlets to augment her teaching, and she performed editorial duties for the Christian Science Publishing Society.[72] She had come a long way since the little two-room schoolhouse in the Michigan frontier town. She also moved to England, where she continued teaching and practicing spiritual healing. Apparently, however, her belief in direct observation and self-discovery—the same values that inspired the nature-study idea of her early student and which she thought underlay the Christian Science movement—proved to her ultimate detriment among the leadership of the Church of Christ, Scientist. In 1902, a series of charges were brought against Julia Field-King by fellow members of the church for straying in her teachings from the dogmas laid out in Mary Baker Eddy's book *Science and Health with Key to the Scriptures*. She believed that Eddy's text left room for new spiritual insights—her accusers labeled *her* insights as heretical—and she disagreed with some of Eddy's published statements: she was effectively excommunicated from the church.[73]

In that way, the organization in which Julia Field-King finally rose to prominence ultimately disowned her. Through all of her remarkable life's turns, however, she continued to think of herself as a teacher. She was still living in England when Bailey reconnected with her after those many years in 1909, and he kept the letter she sent him in response, neatly handwritten in flowing cursive, which still sits in the archives of Cornell University and serves as a testament to the enduring legacy that visionary teachers everywhere leave with their pupils—as well as the enduring imprint that pupils make on their teachers in turn. It also beautifully illustrates the unexpected delight of a teacher hearing from a student many years after having taught him and recognizing that she had made a real difference in his life. Her life's many adventures had not dimmed her memory of him, it seems, or his family. Her letter came just a month before the revised edition was published with its dedication to her, in the fall following Bailey's attempt to leave the deanship at Cornell and take up a full-time arrangement with Macmillan. It is not clear whether Bailey ever told her that he dedicated the book to her.

Sept 14[,] 1909.

How shall I address you? I want to say 'My dear boy';—but when I look at the fine face that the likeness you so kindly sent, which so surely indicates the strong, yet gentle, *man* whose faithfulness to high ideals has given him such a high place in the noble calling of teacher, I hardly dare say it. But then the true man never loses quite the heart of the boy who is "father to the man." It was very sweet to read your, "My dear teacher!" As I look back I can see in the light of the added years of experience, how much I fell short of the best wisdom in the work of a real teacher. Were I now 30 years old with what life has taught me, I would choose still to be a teacher—better fitted to attain to a higher ideal. [. . .]

I am so glad to hear about your dear wife and lovely daughters. How wise you are to equip them (your daughters) so royally to meet the exigencies of life. How much more strenuous the simplest life to-day seems to be than fifty or seventy-five years ago! I wish I might know all your dear ones. I cannot tell you how near you seem to me. It is all so sweet and charming to hear from one of *my boys*. You know how I loved them. You, yourself, in your office of teacher know what an unselfish, tender love grows in one's heart for the lads and lasses who come to us for help and guidance. It is a more unselfish and impersonal affection than any other relation in life inspires. [. . .]

Of course you and Mrs. Bailey will call at this port on your way around the world. I shall look forward with great pleasure to having the joy of meeting you and the honour of entertaining you, to the best of my ability. Would it be asking too much for you to send me a postcard from the different places you visit on your tour? I know how almost impossible letters are when traveling. Do not send even a postcard if it is a care or trouble. I do hope your travels will bring you to my door soon. A warm welcome will await you both. Give my love to dear Mrs. Bailey and your father & mother.

Very lovingly yours,

Julia Field-King

"The power that moves the world," as Bailey writes, continues to be "the power of the teacher." At a time in which teachers find themselves under fire, too often scapegoated by political opportunists and weighed down by forces that seek to over-professionalize children, teachers remain among the most significant and most undercompensated workers in our society. Strong teachers continue to understand intuitively the benefits of education out-of-doors, and nature-study lives on, even as terminology changes and falls away. The power of the teacher to foster in her students "a living sympathy with everything that is" continues to hold forth the promise of a better world, tying the individual to the larger biotic community, developing the call to fellow service and strengthening the ties of democratic union, building a world of curious, sensitive, and caring local/global citizens. The foundation of a thriving future planet lies in this "outlook to nature," and the path to such understanding, wherever we find it and whatever teachers lead us along the way, is nature-study.

NOTE ON THE TEXT

No manuscript for *The Nature-Study Idea* is known to exist. The book went through at least four editions and numerous printings, first with Doubleday and later with Macmillan. After the first edition in 1903, a revision came between the 1904 and 1905 printings, but the edits were relatively minor, and the 1905 printing was not labeled a new edition. Bailey's name, in an anomaly among his books, appeared on the outside cover of the Doubleday editions and in advertisements for the book as "Liberty H. Bailey," although on the cover page it appears as it did on nearly all his books: "L. H. Bailey." The first Macmillan printing in 1909 marks the one major revision of the text, in which large sections were cut and moved around and much new text was incorporated, especially to Part III. Bailey's deep engagement is clear throughout that revision, which shows a refinement of both thought and language while remaining true to the spirit of the original. The text was completely reset, and it was printed in sextodecimo (the Doubleday edition was closer to an octavo) to match the format of the other books that would become, with it, The Rural Outlook

Set. It was labeled on the title page "Third Edition, Revised," with "Revised Edition" in gilt lettering on the lower spine. A "Fourth Edition" was then published in 1911, the same year that *The Country-Life Movement in the United States* was also published, with another set of mostly minor revisions. It was the first edition of *The Nature-Study Idea* in which the list of titles in the Rural Outlook Set was printed facing the title page, as reproduced in this edition, below.

The present text is based on the 1920 printing of the fourth edition, the latest printing during Bailey's lifetime that I have been able to locate, knowing that Bailey sometimes made changes silently between printings of his other books.[1] Aside from cheaply produced print-on-demand copies that simply reproduce scans of the old editions, and one recent thrift edition in 2021 that contains numerous typographical errors (even in chapter titles), the book appears to have been out of print since 1920. A full apparatus comparing all editions of the work has not been attempted here, as the aim of this volume is primarily to reintroduce the work, and such an apparatus would necessarily become voluminous and ungainly. However, many beautiful and interesting passages were completely cut for the third edition, and in the interest of giving the reader as full a text of *The Nature-Study Idea* as possible, I have endeavored to provide all substantive cuts made for the third edition, as they appeared in the first edition: brief passages appear in the endnotes and several lengthy sections appear following the text under the heading Major Sections Restored from the First Edition. Wherever Bailey transposed a passage to another place in the text, made cuts simply to eliminate redundancy or to smooth out a passage without losing meaning or content, or made other minor corrections of a stylistic or typographical nature, his final revision is left to stand without comment as it appears in the final 1920 printing. But those readers of the first edition who might otherwise miss passages such as Bailey's poem, "Child with the gray-blue eyes," or the lyrical concision of the "Conclusion" chapter of Part I (some of which was rearranged and incorporated into the first chapter of Part II), will find all of that cut material in the present volume. Moreover, the near-complete rewrite of the chapter that would become "Nature-Study Agriculture" in the Macmillan editions illuminates an evolution of thinking and a historical shifting of circumstances that will be of interest to careful readers. The first-edition text of that chapter, included here under Major Sections Restored, delineates the early nature-study work at Cornell in much greater detail than does the text of the third and fourth editions, and it names more of the educators who were central to the program's evolution and leadership, so it is of special historical interest. It is my hope that, with all these inclusions, this text will best serve to introduce the full work to new generations of readers and that it also might best meet the needs of those seasoned Baileyateurs who may not have known that their beloved old 1905 printing had ever come under the author's own ruthless editorial pen.

Footnotes are Bailey's unless otherwise noted, and editorial commentary has generally been restricted to endnotes. Throughout this volume, square brackets within quotations and footnotes indicate editorial insertions, and bracketed ellipses are used to distinguish from ellipses in the original. Where square brackets appear in the main text of *The Nature-Study Idea* and in the Related Writings, they belong to Bailey. Parenthetical cross-references within the text of *The Nature-Study Idea* have been rekeyed to this print edition.

I have endeavored, in the selection of book reviews of the first and third editions of *The Nature-Study Idea*, to present a range of perspectives from respected periodicals, ranging from literary critics to scholars of education and sociology. In each case I have excerpted the full text as it pertains to the book, excising material devoted to other works when necessary and citing my source. Then, a section titled Related Writings presents a series of essays, written by Bailey or in direct response to him and published primarily in periodicals, that trace the evolution of Bailey's nature-study idea from the early years of Cornell's nature-study work in 1897 through the time that Bailey made his major revision to the book in 1908–1909. Further discussion of those writings appears in the Note on the Selections at the beginning of that section.

The Nature-Study Idea

The Rural Outlook Set

✠

The Outlook to Nature (revised)
The Nature-Study Idea
The State and the Farmer
The Country-Life Movement

THE NATURE-STUDY IDEA

An Interpretation of the New School-Movement to Put the Young into Relation and Sympathy with Nature

L. H. BAILEY

TO
Mrs. Julia Field-King
A TEACHER WHO ALLOWED
A BOY TO GROW
I inscribe this book

Contents

Part I

Being an attempt to define and explain what nature-study is

I

WHAT IS NATURE-STUDY?

A CONTRIBUTOR to a recent issue of a leading technical journal has endeavored to find a satisfactory answer to the question, "What is nature-study?" by appealing to "eminent scientific men."[1] The answers of these men are printed there in full.

Now, the nature-study movement is not a product of "eminent scientific men," nor directly of the current natural-science movement. It is a product of the common schools.[2] Eminent scientific attainment, as such, is not to be expected to enable persons to give satisfactory answer to the question, for the subject is not within its realm. Happily, many scientific men are also closely in touch with elementary education, and therefore are fully competent to discuss the nature-study movement, but it is this very touch with the common schools, not their eminent scientific achievements, that gives them this competency; and some of the answers referred to above are good definitions from the child-teacher's point of view.

To be sure, the term nature-study etymologically implies only the study of nature; and "nature" is conventionally understood to mean the world of outdoor objects and phenomena. But all words and terms mean less or more than their mere etymology would imply, and this meaning is determined by usage. So usage has determined a definite office for the name nature-study: it designates the movement originating in the common schools to open the pupil's mind by direct observation to a knowledge and love of the common things and experiences in the child's life and environment. It is a pedagogical term, not a scientific term.

Nature-study is not synonymous with the old term "natural history," nor with "biology," nor with "elementary science." It is not "popular science." It is not the study of nature merely. Nature may be studied with either of two objects: to discover new truth for the purpose of increasing the sum of human knowledge; or to put the pupil in a sympathetic attitude toward nature for the purpose of increasing his joy of living. The first object, whether pursued in a technical or elementary way, is a science-teaching movement, and its professed purpose is to make investigators and specialists. The second object is a nature-study movement, and its purpose is to enable every person to live a richer life, whatever his business or profession may be.

Nature-study is a revolt from the teaching of formal science in the elementary grades. In teaching-practice, the work and the methods of the two intergrade, to be sure, and as the high-school and college are approached, nature-study passes into science-teaching, or gives way to it; but the intentions or motives are distinct—they should be contrasted rather than compared. The nature-study method is a fundamental and, therefore, a general educational process; the formal science-teaching method is adapted to mature persons and to those who would know a particular science.

Nature-study, then, is not science. It is not knowledge. It is not facts. It is spirit. It is an attitude of mind. It concerns itself with the child's outlook on the world.

Nature-study will endure, because it is natural and of universal application. Methods will change and will fall into disrepute; its name will be dropped from courses of study; here and there it will be incased in the schoolmaster's "method" and its life will be smothered; now and then it will be over-exploited; with some persons it will be a fad: but the spirit will live.

So common is the misconception of the meaning and mission of the nature-study movement, that I cannot resist the temptation to bring together in book form a few notes and essays on some of the more salient features of it, even if the resulting book lack somewhat in homogeneity and have some repetitions. These pieces have been written at intervals in the past six years. Most of them were prepared for specific occasions, for the purpose of discussing disputed points or of answering challenges; some have been written specially for this collection. Some of them have been published. They are offered in all humbleness, since every person's view is necessarily colored by his own field of work; but on the main thesis—that nature-study teaching is one thing and that science-teaching for science's sake is another—I have no hesitation.

The foregoing paragraph indicates the make-up of the original edition of this book, which was published by Doubleday, Page & Co. in 1903. The book appears to have found a constituency beyond my expectations, and the continued use of it influences me now (1909) to make a new edition.[3] If I were writing the book anew at this time, I might put it in different phrase; but as it was written when I was actually engaged in teaching and was filled with the practical details of the subject, and as so many parts of it have been so often quoted, I shall leave it much as it was originally prepared. Since the book was written, I have ceased all teaching and have been consumed in educational administrative work.[4] I have therefore seen the subject from a different angle; but on going over the text I find nothing that I would change in the fundamental contentions. In fact, I have a deeper conviction than ever that the method and point of view of the nature-study people are bound to exercise great influence in redirecting our education.

I have a growing feeling that the nature-study method is not only a public-school process, but that it is equally needed in colleges and universities for all unspecialized students. The process applies, in fact, from kindergarten to college. From long experience I am convinced that much of our college physics, botany, zoölogy and chemistry is very poorly taught if we are to consider its effect on the student; and this effect is, of course, the end of teaching. A student may take college physics and yet have little conception of the common physical phenomena of life. He may study physiology and gain little real understanding of his bodily functions or of every-day sanitation. These subjects are likely to be taught with the special student in mind rather than the general student. The teacher is disposed to think of the necessity of developing a whole subject rather than to give the student a rational and vivid conception of the material as it relates to him. I have been interested all my life in plants; but I should not care to have one of my pupils devote four or five periods a week for a whole freshman year to the study of botany unless he were specially interested in botany. Much of the beginning teaching in the sciences in colleges and universities is undoubtedly very bad. It is no doubt accurate, and it may also be adapted to the few students who desire to specialize in the subject; but such students should be taken further in courses designed for them. Condensed general courses that give the college student a rational view of the subject, without many details and exceptions, are very much to be desired; and such courses should attempt to relate the student to his own experience in life.

We have been passing through a long epoch of speech-education. This no doubt is largely the outcome of the results of the Reformation, to teach persons to read their own scripture.[5] The schools must undergo a continual process of growth and adaptation if they are to meet the needs of the passing generations of men. We now feel that speech-education is not a primary educational process, but that real education should grow out of or result from the common activities of the child. Some day we shall set all our children at work when they go to school and make them to be effective men and women in the common work of men and women.

After all these years of nature-study enterprise, it is naturally assumed by many persons that we ought to be able to give statistics of the number of pupils who are enrolled in the subject, the number of teachers that are teaching it, the number of books that have been read, and other exact figures.[6] This supposition misses the very purpose of the nature-study movement, which is to set pupils at work informally and personally with the objects, the affairs and phenomena with which they are in daily contact. There are very many teachers and very many schools, and very many pupils, who have a new outlook on life as the result of nature-study work; but if I could give a statistical measure of the nature-study movement, I should consider the work to have been a failure, however large the figures might be.

The seed has been planted, and it has germinated. The evolution of a new intention in education is under way and is beginning to be felt. The principles have been stated; the current discussions are of methods, difficulties, and of local and personal adaptations.

We are to open the child's mind to his natural existence, develop his sense of responsibility and of self-dependence, train him to respect the resources of the earth, teach him the obligations of citizenship, interest him sympathetically in the occupations of men, quicken his relations to human life in general, and touch his imagination with the spiritual forces of the world.

If life is worth living it must be invigorated, and there is no invigoration without enthusiasm and spirit. We must all have practice in the common affairs of life; but practice alone is dead, and worse than dead. If we cannot add the spirit and the true sentiment to life, then there is no interest in living excepting for that which is gross. It is better to have a thread of inspiring philosophy running through the day's work than to have a very large bank account. This means that a school should have a soul.

The reader will understand that I have approached my subject from the side of fact and of experience, not from the side of pedagogical theory or of the psychology of education.[7] Nature-study is experience-teaching. In my first work and writing on nature-study, I think that I was wholly unconscious of any conflict of my views with the current theories of educational procedure; in fact, the pedagogical theories were unknown to me till they were called to my attention. I had merely set forth my convictions, resulting from many years of teaching, to the effect that the best way to teach nature subjects is to begin with good simple observation rather than with dissection, classification, theorizing or memorizing. I think that the same process should be followed in the training of the teacher himself. I doubt whether saturation in the psychology of pedagogy affords a good start for the training of a teacher. I observe an indefiniteness and haziness of ideas in persons who have their theory before they have their facts. They do not have their feet on the ground. They do not drive stakes; or if they do, they ponder the method until the operation becomes lifeless. For nature subjects, the first essential is an intense love of nature; the best training is to acquire the actual facts and to know the subject, and then to go out and teach, without too much burden of doubt as to the kind and propriety of the theoretical methods. I do not doubt the value of the psychological study of education, and all teachers should profit by a discussion of educational history and method; but we should be careful not to fill the young teacher full of abstractions. A teacher may safely theorize and speculate after he has learned how to teach.

Of the criticisms on this book and on my general attitude toward nature-study teaching, the most important is that I insist too much on spontaneousness and informality and thereby provide an excuse for lazy or indifferent teachers who do not want to make preparation for their lessons. The lazy teacher can find plenty of excuses. One who fairly reads the book need not be misled. My general plea is a challenge to existing hard-and-fast methods and to those ways of teaching that take the pupil prematurely beyond his depth. There is no danger that the school work will lack in formality: our systems encourage formality, and the desire to standardize all methods seems to be extending, but a free and natural procedure needs always to be promoted and defended. In actual school practice, it is of course necessary that a system be followed and that the teacher have ability enough and knowledge enough to be able to teach. I have not cared to prepare an outline for class work: the book is concerned with the nature-study idea. Nor have I desired to make supplemental statements in these intervening years, for I have wanted the idea to sink in.[8]

The recent years have been a time of widespread discussion of all phases of education for the people, and the nature-study idea has received its full share of attention. Whatever may be the opinion of individual teachers and writers on the nature-study movement, it is a fact that our educational methods are re-shaping themselves in such a way as to allow the pupil to develop a sympathetic and vital contact with his usual environment; and the stiff, dead and painfully exact teaching of rule and fact to the young is rapidly giving way to a free, spirited and natural way of teaching. We can even now begin to see the result in a less restrained and more wholesome outlook on life in the young generation. It will be much satisfaction to me if I can feel at the end that this fragmentary book has had some effect in heartening teachers not to be afraid to teach.

II

Who First Used the Term Nature-Study?

A BRIEF history of the origin of the contemporary nature-study movement will clarify our ideas as to its spirit and purpose.[9] I am aware that the history that follows is incomplete, and that persons who were connected with the beginnings of it are not mentioned; but I think that the account will be useful in giving us perspective, and in establishing an approximate date for the first use of the term.

I have engaged in a large correspondence for the purpose of discovering something of the history of the nature-study movement in North America. Oftenest, perhaps, I have been referred to the teaching of Agassiz at Penikese as the beginning, at least in this country.[10] Agassiz, however, did not teach nature-study in the school sense in which we use the term, although he gave us the motto, "Study nature, not books." He taught the study of nature by the "natural method." His instruction was given from the investigator's or the specialist's viewpoint, and it was intended primarily for students and adults.

The present nature-study movement, as I have said, is a product of the elementary schools, not of universities, although many university and college men have been instrumental in forwarding it. Cornell was perhaps the first university to take it up as a distinct enterprise (1895),[11] but the movement was already well under way in many places at that time. At this institution it became an extension-teaching movement. Professor C. F. Hodge of Clark University, under the inspiration of Stanley Hall, began popular work in nature-study in 1897.[12] The Cornell work is not so much a school enterprise as a movement to make use of the schools to reach the people on the farms. This work, more than any other perhaps, has emphasized the nature-sympathy and the nature-relations.

The beginnings of nature-teaching are certainly as old as the time of Socrates and Aristotle. It is concretely expressed in the work of the great educational reformers—Comenius, Pestalozzi, Jean J. Rousseau, Froebel and the others.[13] In a large measure, the spirit of our present-day nature-study movement—which seems so new to us—is a recrudescence. Just now it represents a reaction from the dry-as-dust science-teaching.

What we may legitimately call nature-study, in the current acceptation of the term, began to take form in this country from 1884 to 1890. Who first used the term I do not know; and it is of small consequence, because the term may mean much or nothing. The term appears to have been at first a substitute for "object lessons," "plant work," "elementary science," and the like. Dr. Piez, of the Oswego (N. Y.) Normal School,[14] makes the following comment on the pedagogical origin of the nature-study idea: "I have come to the conclusion that nature-study in spirit, if not in name, is the direct descendant of object teaching. Object teaching aimed at the use of the senses in acquiring knowledge, and was introduced to displace the mechanical 'memory' method[15] current in the schools. It was responsible for raising the problem of method among thoughtful teachers. But the 'lessons on objects' were justly deserving the criticism that they were disconnected, and that the knowledge resulting from them was a knowledge of isolated facts not organized into a comprehensive whole."

Although the teaching of Agassiz may not have been nature-study, as we understand the term, it is undoubtedly true that the present nature-study movement is a proximate result of the forces that he and his contemporaries set in motion. A strong application of this influence to school life was made in Boston by Alpheus Hyatt and Lucretia Crocker.[16] In various places, others of Agassiz's followers carried his spirit into the schools. One of the most powerful early adaptations of his teaching to the common-school work was made at the State Normal School at Oswego, N. Y. There was a strong Pestalozzian influence in this institution, under the leadership of the late Dr. Sheldon.[17] Professor H. H. Straight went to Oswego in 1876. He had come under the influence of Agassiz and Shaler.[18] He was a student of science, but his views of science-teaching in the elementary school underwent gradual but decided change under the Pestalozzian influence in which he was placed. He saw the insufficiency of "object teaching" as an educational process. The defects he sought to overcome by "correlation of the subjects of study." As director of the practice school, he worked out his ideas of correlation in "nature" subjects and geography subjects. His work included the study of the common things in the neighborhood. In 1883 Professor Straight went to the Cook County (Ill.) Normal School and taught there until his death, in 1886. He had great influence in developing the ideals of this institution, and was given credit therefor by Colonel Parker, the distinguished head of the school.[19] So far as I know, however, Professor Straight did not use the term "nature-study."

The introduction of elementary science as an organic part of school work, ranking with arithmetic and grammar, was made in the Cook County (Ill.) Normal School as early as 1889, under the presidency of Francis W. Parker. This introduction was made by the late Wilbur S. Jackman, whose teaching and writing in nature-study lines are well known.[20] In 1884 Mr. Jackman began teaching biology in the Pittsburg High School. During five years' connection with that school he became strongly impressed with the necessity of having a broad foundation laid in the elementary grades for the study of science. The pupils were ignorant of the simplest phenomena that occurred about them. In the spring of 1889 he planned a general course in nature-study and presented it to the superintendent and the principals of the ward schools in Pittsburg. It was agreed that in the fall he should have the privilege of meeting the teachers for the purpose of starting this work in the primary and grammar grades. Before the year closed, however, he received an invitation from Colonel Parker to enter the Cook County Normal School and take up the work with him. He entered on the work in the Cook County Normal School in the fall of 1889. During this year (1889) he elaborated the plan already begun, as above outlined. The features which perhaps most distinguished this scheme of nature-study were: (1) That it adopted the apparently irregular plan of using all the material which the "Rolling Year," season by season, brought into the lives of the children; (2) that it rejected the idea of close and specialized study of inert or dead form and sought to place the children in the fields and woods that they might study all nature at work; and (3) that, instead of looking upon nature-study as being supplementary to reading, writing and other forms of expression, nature-study in itself became a demand that these subjects should be taught. In the fall of 1890 he published bi-monthly pamphlets averaging about seventy-five pages each, which were called "Outlines in Elementary Science." In the spring of 1891, upon the completion of the series, Henry Holt & Company asked the privilege of reprinting and issuing them in book form. This was accomplished. There was considerable correspondence concerning the name, which resulted finally in the adoption of the term "Nature-Study for Common Schools," and this term has been used continuously ever since.

Another, and an independent, movement started nearly simultaneously in Massachusetts, under the leadership of Arthur C. Boyden, now Vice-Principal of the State Normal School at Bridgewater, Mass.[21] In 1889 a committee was appointed in the Plymouth County Teachers' Association to recommend a plan of introducing nature-study into the schools of the county. For a number of years previous to this time a definite series of lessons on minerals, plants and animals had been taught in the Bridgewater Normal School, and many superintendents and teachers who graduated from the school were teaching the subjects in various parts of the county. It seemed to be the time for a concerted plan of work, and a few persons who were interested in it took this means of starting. An outline for the study of trees was prepared and sent to every school in the county, with provisions for a report from each town at the next annual meeting. This plan was continued for a number of years, and usually an exhibition of the results was made. The work secured such a good hold that the committee was finally discontinued. In the same year the subject was taught in the institutes, held each fall and spring throughout the State under the auspices of the State Board of Education, and then for ten years Mr. Boyden taught and lectured in these institutes from one end of the State to the other. Printed outlines and illustrated lessons were given. In 1889, also, a department of nature-study was established in the summer school at Cottage City, and Mr. Boyden carried it till 1901. The definite beginning of the movement, as such, in Massachusetts seems to have been in 1889. At first the work was called "elementary science," but this seemed to be inappropriate, and "nature-study" was suggested. This term seemed to be a good equivalent of the German "naturkunde"—nature knowledge. On all programs it was thus printed and quickly secured standing. Shortly after the movement began, the "Conference of Educational Workers" was established. One of the committees had charge of nature-study and met monthly in Boston. Mr. G. H. Martin, Agent of the Board of Education, was chairman, and Mr. Boyden was secretary.[22] They worked out courses of study for distribution, and one year they had a large exhibit from the whole State of the results of the work. These exhibits were common in cities between 1890 and 1895.

Amos M. Kellogg, editor of the "New York School Journal" from 1874 to 1904, was one of the early writers and advocates on the necessity of drawing on the world about us in the education of the young.[23] Visiting a school in Monroe County, Pennsylvania, in 1885, where the teacher was imbued with enthusiasm in this direction and asked for special directions, he suggested to Frank Owen Payne (who was then a regular contributor to the "School Journal"), the preparation of specific lessons;[24] as the term nature-study came to be used he suggested to Mr. Payne the need of the hyphen between the words, and this came to be in regular employment.[25] The specific lessons prepared by Mr. Payne took the title of "One Hundred Lessons Around the School." Mr. Payne began the employment of practical nature-study in 1884 when a teacher at Corry, Pennsylvania; then in 1885–86 in New Jersey. He lectured on the subject in Minnesota in 1886–89, and has written on it for educational journals.

Many schools in several states were introducing elementary science in the latter part of the eighties, and it seems that several of them began to use the word nature-study without knowing where or how the term was suggested. The term is now in widespread use in English-speaking countries.

The word nature-study was used in January, 1905, in the title of a monthly magazine, "The Nature-Study Review," edited and published by Professor M. A. Bigelow of Teachers College, Columbia University, with a board of advisory editors.[26] In January, 1908, the "American Nature-Study Society" was organized, and the Review is now its official organ.[27]

III

The Meaning of the Nature-Study Movement

IT is one of the marks of the progress of the race that we are coming more and more into sympathy with the natural world in which we dwell. The objects and phenomena become a part of our lives. They are central to our thoughts. The happiest life has the greatest number of points of contact with the world, and it has the deepest sympathy with everything that is.[28]

The best thing in life is sentiment; and the best sentiment is that which is born of the most accurate knowledge. I like to make this application of Emerson's injunction to "hitch your wagon to a star"; but it must not be forgotten that a person must have the wagon before he has the star, and he must take due care to stay in the wagon when he rides in space.[29] Mere facts are dead, but the meaning of the facts is life. The getting of information is but the beginning of education. "With all thy getting, get understanding."[30]

Of late years there has been a rapidly growing feeling that we must live closer to nature and make our nature-sentiment vital; and we must of course begin with the child. We attempt to teach this nature-love in the schools, and we call the effort nature-study. It would be better if it were called nature-sympathy.

As yet there are no recognized and regulated methods of teaching nature-study. The subject is not a formal part of the course of study; and thereby it is not perfunctory. And herein lies much of its value—in the fact that it cannot be reduced to a mere system, is not cut and dried, cannot become a part of rigid and formal school method. Its very essence is spirit. It is as free as its subject-matter, as far removed from the museum and the cabinet as the living animal is from the skeleton.

It thus transpires that there is much confusion as to what nature-study is, because of the different attitudes of its various exponents; but these different attitudes are largely the reflections of different personalities and the working out of different methods. We cannot say that one way is right and another wrong. There may be twenty best ways of teaching nature-study. The mode is essentially the expression of one's outlook on the world.[31] Heretofore, we have put the emphasis on training for heaven and taking the child out of his world.[32]

The reader who has followed me thus far has got at the kernel of my thought. I shall now go into more detail, with the purpose to relate the discussion to the practical work of the schoolroom, to develop the teacher's attitude, and to state the essential nature of the movement in different ways and from different angles in order that the thought may stick. This chapter, therefore, is a budget of suggestions rather than an analysis.

What nature-study is not

There are two or three fundamental misconceptions of what nature-study is or should be; and to these we may now give attention.

It is not the teaching of science—not the systematic pursuit of a logical body of principles. Its intention is to broaden the child's horizon, not primarily to teach him how to widen the boundaries of human knowledge. It is not the teaching of botany or entomology or geology, but of plants, insects and fields. But many persons who are teaching under the name of nature-study are merely teaching and interpreting elementary science. Fundamentally, nature-study is seeing what one looks at and drawing proper conclusions from what one sees; and thereby the learner comes into personal relation with the object.

It is not reading from nature-books. Nature-study is studying things and the reason of things, not about things. A child was asked if she had ever seen the great dipper. "Oh, yes," she replied, "I saw it in my geography." This is better than not to have seen it at all; but the proper place to have seen it is in the heavens. Nature-readers may be of the greatest value if they are made incidental and secondary features of the instruction; but, however good they may be, their influence is pernicious if they are made to be primary agents. Nature-study begins with the concrete, as the child does if left to itself.[33] The child should first see the thing. It should then reason about it. Having a concrete impression, it may then go to the book to widen its knowledge and sympathies. Having seen mimicry in the eggs of the aphis on the willow or apple twig, or in the walking-stick, the pupil may then take an excursion with Wallace or Bates to the tropics and there see the striking mimicries of the leaf-like insects. Having seen the wearing away of the boulder or the ledge, he may go to Switzerland with Lubbock and see the mighty erosion of the Alps.[34] Now and then the order may be reversed with profit, but this should be the exception: from the wagon to the star should be the rule.

Nature-study is not the teaching of facts merely for the sake of the facts, or materials for the sake of the materials: its purpose is to develop certain intellectual powers by the use of the materials. It is not the giving of information only—notwithstanding the fact that some nature-study leaflets are information leaflets.[35] We must begin with the fact, to be sure, but the lesson lies in the significance of the fact. It is not necessary that the fact have direct practical application to the daily life, for the purpose is the effort to train the mind and the sympathies and to develop in the child a correct view of nature. It is a common notion that when the subject-matter is insects, the pupil should be taught the life-histories of injurious insects and how to destroy the pests. Now, nature-study may be equally valuable to the pupil, whether the subject is the codlin-moth or the ant, since both may be within his sphere and his relations; but to confine the pupil's attention to insects that are injurious to man is to give him a distorted, partial and untrue view of nature. A bouquet of daisies does not represent a meadow.

It is not a program for the teaching of morals. Children should be interested more in seeing things live and in studying their habits than in killing them. Yet I should not emphasize the injunction, "Thou shalt not kill."[36] I should prefer to have the child become so much interested in living things that it would have no desire to kill them. The gun and sling-shot and steel-trap will be laid aside because the child does not care for them any more. We have been taught that one must make collections if he is to be a naturalist; but collections alone make museums, not naturalists. The scientist needs these collections; but it does not follow that children always need stuffed animals, birds' eggs, and bottled specimens, although it is important to encourage a regulated collecting instinct.

Nature-study is not merely the adding of one more thing to a course of study. It is not coördinate with geography or reading or arithmetic. Neither is it a mere accessory, or a sentiment, or an entertainment, or a means of injecting vacant wonder into the pupils. It is not "a study." It is not the addition of more "work." A new "study" taught by the old method would not represent progress. The idea has to do with the whole point of view of elementary education, and therefore is underlying. It is the full expression of personality. It relates schooling to living. It is a practical working out of the extension idea that has been so much a part of our time.[37] More than any other recent movement, it will reach the masses and revive them.

Nature-study should not be unrelated to the child's life and circumstances. It stands for directness and naturalness. It is astonishing, when one comes to think of it, how indirect and how remote from the lives of pupils much of our education has been. Geography still often begins with the universe, and finally, perhaps, comes down to some concrete and familiar object or scene that the pupil can understand. Arithmetic has to do with brokerage and partnerships and partial payments and other things that mean nothing to the child. Botany begins with cells and protoplasm and cryptogams. History deals with political and military affairs, and only rarely comes down to physical facts and to those events that express the real lives of the people; and yet political and social affairs are only the results or expressions of the way in which people live. Readers begin with mere literature or with stories of scenes the child will never see. Of course these statements are meant to be only general, as illustrating what is even yet a great fault in educational methods. There are many exceptions, and these are becoming commoner. Surely, the best education is that which begins with the materials at hand. A child knows a stone before it knows the earth.

The outlook by fact and by fancy

There are two ways of interpreting nature—the way of fact and the way of fancy. To the scientist and to the average man the interpretation by fact is usually the only admissible one. He may not be open to argument or conviction that there can be any other truthful way of knowing the external world. Yet, the artist and the poet know this world, and they do not know it by cold knowledge or by analysis. It appeals to them in its moods. Yet it is as real to them as to the analyst. Too much are we of this generation tied to mere phenomena.

We have a right to a poetic interpretation of nature.[38] The child interprets nature and the world through imagination and feeling and sympathy. Note the intent and sympathetic face as the child watches the ant carrying its grains of sand and pictures to itself the home and the bed and the kitchen and the sisters and the school that comprise the ant's life. What does the flower think? Who are the little people that teeter and swing in the sunbeam? What is the brook saying as it rolls over the pebbles? Why is the wind so sorrowful as it moans on the house-corners in the dull November days? There are elves whispering in the trees, and there are chariots of fire rolling on the long, low clouds at twilight. Wherever it may look, the young mind is impressed with the mystery of the unknown. The child looks out to nature with great eyes of wonder.[39]

We cannot say that the good poets have not known nature, because they have not interpreted by fact alone. Have they not left us the essence and flavor of the fields and the woods and the sky? And yet they were not scientists. So different are these types of interpretation that we all unconsciously set the poet over against the scientist.

Good poetry is not mere vacant sentiment. The poet has first known the fact. His poetry is misleading if his observations are wrong. Whatever else we are, we must have the desire to be definite and accurate. We begin on the earth; later, we may drive our Pegasus to a star.[40]

Of course I would not teach nature-subjects in order that the poetic point of view may be enforced. I plead only that the poetic interpretation is allowable. It may be one result of knowing nature for the sake of knowing it.

How nature-study may be taught

How shall nature-study be taught? By the teacher and the object. The teacher will need helps. There are books and leaflets that will help him. These publications may be put in the hands of pupils if it is always made plain that the recitation is to be from objects and situations that the pupil has seen, not from the book. There can be no text-book of real nature-study, for when one studies a book he does not study nature. The book should be a guide to the animal or plant: the animal or plant should not be a guide to the book.

The teacher may need the help of a program or consecutive purpose. The program, however, should not be a tabulated series of regulations or a hard-and-fast system; but there should be some underlying educational principle or intention running through every item of it. The work may be informal and free without being aimless.

This immediate purpose or plan may be to teach the progress of the seasons; the common implements and simple handcrafts; the plant life of the neighborhood; the bird life; the usual insects; the heavens; the weather and its relations with man and animals; something of the farming or industries of the region; one's own mind and body and how they should be governed in the interest of good health; or some other theme that will tie the work together. In practice, the work will almost necessarily be consecutive because the teacher will feel himself competent in two or three lines and will devote himself to them. The environment will suggest the work.

There will be opportunity for endless variation in the details and in the little applications of the work. The personality of the teacher must always stand out strongly. We need the very best of teachers for nature-study—those who have the greatest personal enthusiasm, and who are least bound by the traditions of the classroom. The teacher, to be ideal, must have more time, more feeling, and more knowledge. It is better if the teacher have a large knowledge of science, but nature-study may be taught without great knowledge if one sees accurately and infers correctly from the particular subject in hand.

The teacher should avoid starting with definitions and the setting of patterns. Definitions should be the result or summary of the study, not the beginning of it. Mere patterns should afford means of comparison only, and not be regarded as useful in themselves; and even then they are often misleading. The old idea of the model flower is an unfortunate one, because the model flower does not exist in nature. The model flower, the complete leaf, and the like, are inferences; and the pupil should not begin with abstract ideas. In other words, the ideas should be suggested by the things, and not the things by the ideas. "Here is a drawing of a model flower," the old method says; "go and find the nearest approach to it." "Go and find me a flower," is the better method, "and let us see what it is."[41]

Two factors determine the proper subjects for any teacher to choose for nature-study instruction. First, the subject must be that in which the teacher is most interested and of which he has the most knowledge; second, it must represent that which is commonest and which can be most easily seen and appreciated by the pupil, and which is nearest and dearest to his life.

With children, begin with naked-eye objects. As the pupil matures and becomes interested, the simple microscope may be introduced now and then. Children of twelve years and more may carry a pocket lens; but the best place to use this lens is in the field. The best nature-study observation is that which is done out-of-doors; but some of it can be made from material brought into the schoolroom.

The tendency is to go too far afield for the subject-matter. We are more likely to know the wonders of China or Brazil than of our own brooks and woods. If the subject-matter is of such kind that the children can see the objects as they come and go from the school, and collect some of them, the results will be the better. As the pupil matures, he should be taken out to the world activities.

It is a sound educational principle that the child should not be taught mere dilutions of science. The young child cannot understand cross-fertilization of flowers, and should not be taught the subject. It is beyond the child's realm. When we teach it to young children, we are only translating what grown-up investigators have discovered by means of faithful search. At best, it will only be an exotic thing to the child. Pollen and stamens are not near and dear to the child.

There are three steps in the teaching of nature-study:

(1) The fact,
(2) The reason for the fact,
(3) The interrogation left in the mind of the pupil.

It is impossible to find a natural-history object from which these three factors cannot be drawn, for every object is a fact and every fact has a cause, and children may be interested in both the fact and the cause. It may be better, of course, to choose definite subjects, taking pains, at least at first, to choose those having emphatic characters.

But even in the dullest days of winter sufficient materials may be found to keep the interest aflame. A twig or a branch may be at hand. There should be enough specimens to supply each child. Let the teacher ask the pupils what they see. The replies will discover the first factor in the teaching— the fact. However, not every fact is significant to the teacher or to the particular pupils. It remains for the teacher to pick out the fact or answer that is most significant. The teacher should know what is significant and he should keep the point clearly before him. One pupil says that the twig is long; another that it is brown; another that it is crooked; another that it is from an apple tree; another that it has several unlike branchlets or parts. Now, this last reply may appeal to the teacher as most significant. Stop the questioning and open the second epoch in the instruction—the reason why no two parts are alike. As before, from the great number of responses the significant reason may be developed: it is because no two parts have lived under exactly the same conditions. One had more room or more sunlight and it grew larger. The third epoch follows naturally: are there any two objects in nature exactly alike? Let the pupils think about it.

Choose a stone. If similar stones are in the hands of the pupils, you ask first for the observation or the fact. One says that the stone is long; another, it is light; another, it is heavy; another, that the edges are rounded. This latter fact is very significant. You stop the observation and ask why it is rounded. Some one replies that it is because it is water-worn. Query: Are all stones in brooks rounded? Numberless applications and suggestions can be made from this simple lesson. What becomes of the particles that are worn away? How has soil been formed? How has the surface of the fields been shaped and molded?

It is not necessary that the teacher always know the reason. He may propose that they all find out and report. It is the strong teacher who can say: "I do not know." If a problem had been sent to Agassiz or Asa Gray and he had not understood it, would he have dissimulated or have evaded in the answer?[42] Would he not have said unhesitatingly, "I do not know"? Such men delve for knowledge, but for every fact that they discover they turn up a dozen mysteries. Knowledge begins in wonder. The consciousness of ignorance is the first result of wonder, and it leads the pupil on and on: it is the spirit of inquiry.

These illustrations are given merely as examples. They may not be ideal, but they show what can be done with very common material. In fact, the surprise and interest is often all the greater because the objects are so very common and familiar.

To my mind, one of the best of all subjects for nature-study is a brook. It affords studies of many kinds. It is near and dear to every child. It is an epitome of the nature in which we live. In miniature, it illustrates the forces that have shaped much of the earth's surface. It reflects the sky. It is kissed by the sun. It is rippled by the wind. The minnows play in the pools. The soft weeds grow in the shallows. The grass and the dandelions lie on its banks. The moss and the fern are sheltered in the nooks. It comes one knows not whence; it flows one knows not whither. It awakens the desire to explore. It is fraught with mysteries. It typifies the flood of life. It "goes on forever."[43]

In other words, the reason why the brook is such a perfect nature-study subject is the fact that it is the central theme in a scene of life. Living things appeal to children.[44] To relate the nature-study work to living animals and plants should constitute the burden of the effort. I would study a brook or a fence-corner or a garden-bed or a bird or a domestic animal or an insect or a plant. The life-histories of certain insects, and all common forms of life, afford excellent nature-study exercise for pupils of proper age.

However, the teacher and the way of teaching are more important than the subject-matter, and there are good nature-study teachers who are better fitted to teach inanimate than animate subjects. There is no better nature-study exercise than to observe the erosion by brooks, floods, and rains, if the teacher is prepared to handle it; and surely nothing can be more important than to put the child in sympathy with the weather; and all persons should have the habit of looking at the heavens in day and night.

It is due to every child that his mind be opened to the voices of nature. The world is always quick with sounds, although our ears are closed to them. Every person hears the loud songs of birds, the sweep of heavy winds and the rush of rapid rivers or the sea; but the small voices with which we live are known not to one in ten thousand. To be able to distinguish the notes of the different birds is one of the choicest resources in life, and it should be one of the first results of a good education. It is but a step from this to the other small voices,—of the insects, the frogs and toads, the mice, the domestic animals, the flow of quiet waters, and the noises of the little winds. It is a great thing when one learns how to listen.[45] At least once, every young person should sleep far out in the open, preferably in a wood or the margin of a wood, that he may know the spirit and the voices of the night and thereafter be free and unafraid.

Similar remarks may be made of the odors, for the world breathes a multitude of fragrances of which most persons are wholly unaware. Usually only the strong smells are known to us, and we merely divide them into two classes,—those that we like and those that we do not like.

All the senses should be so trained and adjusted that all our world becomes alive to us. Then we are really sensitive.

One of the first things that a child should learn when he comes to the study of natural history is the fact that no two objects are alike. This leads to the correlated fact that every animal and plant contends for an opportunity to live, and this is the central theme in the study of living things. The world has a new meaning when this fact is understood. This is the key that unlocks many mysteries, and it is the means of establishing a bond of sympathy between ourselves and the world in which we live.

It is a common mistake to attempt to teach too much at each exercise; and the teacher is also appalled at the amount of information that he must have. Suppose that one teaches two hundred and fifty days in the year. Start out with the determination to drop into the pupils' minds two hundred and fifty suggestions about nature. One suggestion is sufficient for a day. Let them think about it and ponder over it. We stuff our children so full of facts that they cannot digest them. I should prefer ten minutes a day of nature-study to two hours; but I should want it quick, sharp, vivid and spontaneous. I should want it designed to develop the observing and reasoning powers of the child and not to gorge the pupil. Spirit counts for more than knowledge.

It is well to verify observations and conclusions on different days. Let the pupils compare ideas and experiences. This develops an intellectual habit of taking nothing on hearsay or for granted.

Taught in this way, nature-study work is not an additional burden to the teacher, but may be made a relief and a relaxation. It may come at the opening of the school hour, or at the close of a hard period, or at other time when an opportunity offers. It may often be combined with the regular studies of the school, and in that way it may be introduced in places where it would otherwise meet with objection. For example, the subject-matter of the nature-lesson may be used for the exercise in drawing or in geography. Let the child draw the twigs; but always be careful that the drawing does not become more important than the twigs.

My remarks on procedure are meant, of course, to apply to children. As the pupil advances, the work will naturally become more systematic, until, in the high school, it may develop into more formal teaching, and then a regular period will be required. Those who complain that nature-study is desultory are really thinking of science, not of nature-study. Although not the teaching of science, as such, nature-study is not unscientific. It is not in any sense a letting down of standards, if properly handled, but a new intention in education.

What may be the results of nature-study?

Its legitimate result is education—the developing of mental power, the opening of the eyes and the mind, the civilizing of the individual. As with all education, its central purpose is to make the individual happy; for happiness is nothing more nor less than pleasant and efficient thinking, coming from a consciousness of the mastery, or at least the understanding, of the conditions in which we live.[46]

The happiness of the ignorant man is largely of physical pleasures; that of the educated man is of intellectual pleasures. One may find comradeship in a groggery, the other may find it in a dandelion; and inasmuch as there are more dandelions than groggeries (in most communities), the educated man has the greater chance of happiness.

Some persons object to nature-study because it is not systematic and graded. They think that it leads to disjunctive and discursive work. The informality may be its charm. Thereby comes the contrast with the perfunctory school work; and thereby, also, arises its naturalness and its freedom. It is easily possible to "organize" nature-work until it becomes as automatic as other work. The formal school work will supply the drill in method and system. Nature-study will afford relaxation, and it will be valuable because it is short, forceful, and voluntary; and this result is worth securing.[47]

The mode of presentation that naturally develops in nature-study teaching is really very important in its effect on the pupil's approach to subject-matter and on his outlook to the world. The presentation is quick, simple, direct, little confused by apparatus and self-consciousness and side issues.

Good nature-study teaching develops personality and encourages the pupil to think for himself and to maintain an individual relation to his world. It emphasizes adaptation to life as distinguished from the tendency of much of our teaching to produce uniformity of thought and action.

Nature-study not only educates, but it educates nature-ward; and nature is ever our companion, whether we will or no. Even though we are determined to shut ourselves in an office, nature sends her messengers. The light, the dark, the moon, the cloud, the rain, the wind, the falling leaf, the fly, the bouquet, the bird, the cockroach—they are all ours.[48] Few of us can travel. We must know the things at home.

Nature-love tends toward simplicity of living. It tends country-ward. "God made the country."[49]

Nature-study ought to revolutionize the school life, for it is capable of putting new force and enthusiasm into the school and the child. It is new, and therefore is called a whim. A movement is a whim until it succeeds. We shall learn much, and shall outgrow some of our present notions, and shall eliminate the vagaries. It is in much the stage of development that manual-training and kindergarten work were twenty-five years ago. We must take care that it does not crystallize into science-teaching on the one hand, nor fall into mere sentimentalism and gush on the other.[50]

In many ways we are now in a transition period in our school systems. We are living in an era of the material equipment of schools—the erecting of magnificent buildings, the gathering of extensive outfits. This is true of colleges and universities as well as of the common schools. When this era is past, we shall have more money to spend for teachers. Teaching will be a profession requiring better training and commanding more pay, and men teachers will come back to it.

In this evolved and emancipated school, the nature-study spirit will prevail, even though the name itself be lost. This spirit stands for a normal outlook on life. It is the active and creative method. It is a developing of the powers of the pupil, not hearing him recite. In spirit and method, it is opposed to the pouring-in and dipping-out process.[51]

The nature-study effort sets our thinking in the direction of our daily doing. It relates the schoolroom to the life that the child is to lead. It makes the common and familiar affairs seem to be worth the while. It ought to make men and women effective and responsible. Essentially, it is not an ideal for the school any more than it is for the home; but so completely do we delegate all work of teaching and instructing to the school, that nature-study effort comes to be, in practice, a schoolroom subject. The ideal of the parent or the teacher should be to bring the child into natural relations with its world; but whatever may be in the mind and hope of the teacher, so far as the child is concerned the nature-sympathy must come as a natural effect of actual observation and study of definite objects and phenomena.[52]

I will mention two forms of adaptation to life, as illustrations of what I mean. (1) Nature-study teaching ought to utilize, as means of education, the tools that a boy or girl naturally uses. The habits of men are as important as those of other animals. How to use a jack-knife, a hoe, a saw, an auger, a hammer, or other implement by means of which man adapts himself to his conditions, is a very essential part of good teaching, but one that is almost universally neglected. The tools of the household may be made the means of training a girl to a new hold on life. These devices are not to be studied merely as implements, but as a part of the study of the natural history of human beings. All this would constitute a manual-training that would be founded on good sense. (2) The pupil should be taught to make observations on himself. He will find himself to be a very interesting natural-history object. It is just as well to know how a man walks as to know how a horse or a crow walks. The unconscious and automatic habits of men and women are as interesting as those of fish and insects. This kind of observation ought to have remarkable significance to health. It is most strange how little we reason from cause to effect in our own habits of eating and drinking and sleeping and exercise, and how much we rely on the physician to advise us in matters on which we ourselves would be much better judges if we observed ourselves as closely as we observe other objects. The simple regulation of the daily habits of life lies at the foundation of all good health. The application of the nature-study spirit of direct and simple observation of ourselves, with less of the physician's physiology, would benefit the pupil and also our civilization immeasurably.

The great intention of nature-study is to cultivate a sensible interest in the out-of-doors, and to remove all conventional obstacles thereto. Real interest in the out-of-doors does not lie in the physical comfort of being in the open in "good" weather (persons who have this outlook do not know nature), but in spiritual insight and sympathy. One sleeps in the woods or fields not because these are the most comfortable places in which to spend the night, but that he may have communion and freedom.

There is a large public and social result of simple and direct teaching of common things. It explains the relations between man and his environment. It establishes a new sense of our dependence on the natural resources of the earth, and leads us not to abuse nature or to waste our resources. It develops a public intelligence on these matters, and it ought to influence community conduct. All teaching that is direct, native and understandable should greatly influence the bearing of the individual toward his conditions and his fellows, awaken his moral nature, and teach him something of the art of living in the world.[53]

IV

THE INTEGUMENT-MAN

I WROTE a nature-study leaflet on "How a Squash Plant gets out of the Seed."[54] A botanist wrote me that it were a pity to place such an error of statement before the child: it should have read, "How the Squash Plant Gets Out of the Integument."

Of course my friend was correct: the squash plant gets out of an integument. But I was anxious to teach the essence of the squash plant's behavior, not a mere verbal fact—and what child was ever interested in an integument?

It is the old question over again—the question of the point of view and what one is driving at.[55] A person may be so intent on mere literal veracity that he misses the pupil. Much of our natural-science teaching is as hard and dead as the old Latin and mathematics.

It is the fear of the Integument-Man that keeps many a good teacher from teaching nature-study. He is afraid that he will make a mistake in statements of small fact. Now, the person who is afraid of making a mistake is the very person to trust, because he will be careful. Of course he will make mistakes—every one does who really accomplishes anything; but the mistakes will be relatively few: he will at once admit the mistakes and correct them when they are discovered, and the pupils will catch his desire for accuracy and admire the sincerity of his purpose. Pity the man who has never made an error!

The teacher often hesitates to teach nature-study because of lack of technical knowledge of the subject. This is well; but technical knowledge of the subject does not make a good teacher. Expert specialists are so likely to go into mere details and to pursue particular subjects so far, when teaching beginners, as to miss the leading and emphatic points. They are so cognizant of exceptions to every rule that they qualify their statements until the statements have no spirit and no force. There are other ideals than those of dead accuracy. It is more important that any teacher be a good teacher than a good scientist. But being a good scientist ought not to spoil a good teacher.[56] The Integument-Man sees the little things and teaches details, and his teaching is "dry." He lacks imagination.

The child wants things in the large and in relation; when it gets to the high-school or college it may carry analysis and dissection to the limit.

The Integument-Man teaches science, although it is not necessarily the best science. The child wants nature.

The Integument-Man thinks that if any work is only accurate it is thereby of value; and accuracy in nature-study begets accuracy in science, when the pupil takes it up later on. This is all well enough; but the child can be accurate only so far as it can comprehend: it must work in its own sphere; integuments are not in the child's sphere.

The degree of statement is more important than final accuracy—if there is such a thing as final accuracy; all knowledge is relative, and what is within the range of one mind may be far beyond the range of another, and it is folly to try to make the statement as full and accurate for the latter mind as for the former. A very imperfect statement of osmosis is accurate for a child or a young pupil; a fuller statement is accurate for the college student; and a still fuller and exacter statement is accurate for the physicist; but perhaps it is impossible to make any statement of it that is finally accurate. The Integument-Man confuses all these degrees, and thinks that because the statement is inaccurate for the physicist, it is therefore inaccurate for the pupil or the child. Refined verbiage that safeguards the statement to the scientist, may confuse it to the beginner. It may be only pedantry and narrowness. It is not an accident that some of the most useful text-books have been made by persons who do not know too much about the subject.

The Integument-Man is fearful of every word that seems to imply motive or direction in plants and the lower animals. "The roots go here and there in search of food" is wrong because roots do not "go." Seeds do not "travel." Plants do not "prepare" for winter. I wonder, then, whether water "runs" or winds "blow." This verbal preciseness forgets that words are only metaphors and parables, their significance determined by the use of them, and that the essential truth, or the spirit, is what we should search for—expressing it, when found, in language that is alive, unmistakable, and conformed to best usage. We must measure the value of any statement to the child in good part by the strength and vitality of the picture that it raises in the mind (p. 154).[57]

The Integument-Man insists on "methods." The other day a young man wanted me to recommend him as a teacher of one of the sciences in a public school. He explained that he had had a complete course in this and in that; he could teach the whole subject as laid down in the books; he knew all the methods. It was evident that he was well drilled. He had acquired a repertory of facts. These facts were carefully assorted and ticketed, and tucked away in his mental cupboard as embroidered and perfumed napkins are laid away in a drawer. Poor fellow!

Mere details have little educative value. An imperfect method that is adapted to one's use is better than a perfect one that cannot be well used. Some school laboratories are so perfect that they discourage the pupil in taking up investigations when thrown on his own resources. Imperfect equipment often encourages ingenuity and originality. A good teacher is better than all the methods and laboratories and apparatus.

I like the man who has had an incomplete course. A partial view, if truthful, is worth more than a complete course, if lifeless. If the man has acquired power for work, a capability for initiative and investigation, an enthusiasm for the daily life, his incompleteness is his strength. How much there is before him! How eager his eye! How enthusiastic his temper! He is a man with a point of view. This man will see first the large and significant events; he will grasp relationships; he will correlate; later, he will consider the details. He will study the plant before he studies the leaf or germination or the cell. He will discover the bobolink before he looks for its toes. He will care little for mere "methods." His teaching will have freshness.

The Integument-Man is afraid that this popular nature-study will undermine and discourage the teaching of science. Needless to say, the fear is absurdly groundless. Science-teaching is a part of the very fabric of our civilization. All our goings and our comings are adjusted to it.[58] No sane man wishes to cheapen or discourage the teaching of science. Nature-study is not opposed to it. Nature-study prepares the child to receive the science-teaching. Gradually, as the child matures, nature-study may grow into science-learning if the pupil so elect. Science-teaching has more to fear from desiccated science-teaching than it has from nature-study. It is the Integument-Man himself who is discouraging the teaching of science. Everything that is true and worth the while will endure.

All youths love nature. None of them, primarily, loves science. They are interested in the things that they see. By and by they begin to arrange their knowledge and impressions, and thereby to pursue a science. The idea of the science should come late in the educational development of the youth, for the simple reason that science is only a human way of looking at a subject. There is no natural science, but there has arisen a science of natural things. At first the interest in nature is an affair of the heart, and this attitude should never be stifled, much less eliminated. When the interest passes from the heart to the head, nature-love has given way to science. Fortunately, it can always remain an affair also of the heart, but the dry teaching of facts alone tends to divorce the two. When we begin the training of the youth by the teaching of a science we are inverting the natural order. A rigidly graded and systematic body of facts kills nature-study; examinations bury it.

Then teach! If you love nature and have living and accurate knowledge of some small part of it, teach! Do not fear your scientific reputation if you feel the call to teach. Your reputation is not to be made as a geologist or zoölogist or botanist, but as a leader. When beginning to teach birds, think more of the pupil than of ornithology. The pupil's mind and sympathies are to be expanded: the science of ornithology is not to be extended; the science will take care of itself. Remember that spirit is more important than information. The teacher who thinks first of his subject teaches science; he who thinks first of his pupil teaches nature-study. With your whole heart, teach!

Do not be afraid of the Integument-Man.

V

NATURE-STUDY WITH PLANTS

ALL the so-called natural sciences are contributing to the nature-study movement.[59] Plants are so much a part of every landscape, however, we have such constant association with them, and the plant material is so easy to secure, that they afford the very best subjects for nature-study work. One cannot understand the world if he does not know plants.

The methods in plant-study show a distinct development in pedagogical ideas which it may be well to recapitulate. One can make out four fairly well marked stages in the teaching of plant subjects.

First, was the effort to know the names of plants and to classify the kinds. This was a direct reflection of the systematic or classificatory studies of the botanists. The external world had been unknown as to its details, and botanists necessarily attempted inventories of the plant kingdom. Plants must be collected and named. From this impulse arose the herbarium collecting, a method of teaching which was so thoroughly impressed into school methods a generation or two ago that it is still troublesome in many places.[60]

The second stage in plant-study in the American schools was the desire to know the names of the parts of plants. It came with the excellent text-books of Asa Gray and others, in which the results of studies in organography, morphology and histology were organized and defined.[61] These books were nearly as rigid in their systems and methods as text-books of physics; and the pupil recited mostly from the book, with perhaps some accessory observation on plants.

The third epoch is that of training for independent investigation. In very recent times, and chiefly since the death of Gray, the German laboratory methods have been widely copied in America by the many young and painstaking botanists who have studied abroad. As a result there are many high-schools that are equipped with microscopes and apparatus that would have done credit to a college or university a few years ago. The customary laboratory method is a distinct advance on the preceding methods of teaching in the fact that the pupil actually studies plants; but its motive and point of view are distinctly wrong for the elementary school because it attempts primarily to teach botany rather than to educate the pupil. The field of view is also very narrow, and the pupil's mind is likely to be closed to nature and restricted in its range. The stage of the microscope and the tables of the laboratory are poor and narrow ranges for the young mind when there are fields and gardens adjacent. The German laboratory method is no doubt quite perfect for the training of investigators and specialists, but it lacks the inspiration and the educative impulse that young minds need.

The fourth stage is the effort to know the plant as a complete organism living its own life in a natural way. It is marked by a new and vital plant physiology. In the beginning of this epoch we are now living.

Suggestions for plant work

The pupil should come to the study of plants and animals with little more than his natural and native powers. Study with the compound microscope is a specialization to be made when the pupil has had experience and when his judgment and sense of relationships are trained.

A difficulty in the teaching of plants is to determine what are the most profitable topics for consideration. Much of the teaching attempts to go too far and the subjects have no vital connection with the pupil's life. Good botanical teaching for the young is replete with human interest. It is connected with the common associations.

Plants always should be taught by the "laboratory method": that is, the pupil should work out the subjects directly from the specimens themselves; but I should want it understood that the best "laboratory" may be the field, and that the plants are to be studied as plants rather than as dissected pieces.

Specimens mean more to the pupil when he collects them. No matter how commonplace the subject, a specimen will vivify it and fix it in the pupil's mind. A living, growing plant is worth a score of herbarium specimens.

In the secondary schools, botany should be taught for the purpose of bringing the pupil closer to the world with which he lives, of widening his horizon, of intensifying his hold on life. It should begin with familiar plant forms and phenomena.[62] It is often said that the high-school pupil should begin the study of botany with the lowest and simplest forms of life. This is wrong. The microscope is not an introduction to nature. It is said that the physiology of plants can be best understood by beginning with the lower forms. This may be true: but the customary technical plant physiology is not a subject for the beginner. There are better ways of putting the beginner into touch with physiology. The youth is by nature a generalist. He should not be forced to be a specialist.

Just what kind of plant or animal subjects should be taught must depend (1) on the desires and capabilities of the teacher; (2) on the place in which the school is—whether city or country, North or South, prairie or mountain—for it is important that the subject be common and have relation to the experiences of the pupils; (3) on the desires of the pupils, particularly if they are to do the collecting; (4) on the time of the year.

Whenever possible, let the pupil first come into cognizance of the plant as a whole. It is well to choose one species that is common and familiar; then endeavor to determine where it grows, why it grows there, how it is modified in different circumstances. If it is a dandelion, one lesson may be devoted to dandelions in the school-yard; another to dandelions in the meadow; another to dandelions along hard and dry roadsides; another to dandelions in rich farmyards and gardens; another to dandelions in the borders of woodlands. Compare the relative abundance of dandelions in these different places: why? Do the plants "look" the same in these different places: how differ and why? (Note the size and form of plants, relative number of leaves, form and size of leaves, root habit, abundance of bloom, length of flower stems.) It is a practice in some schools to teach mathematics by means of dandelions, on the mistaken notion that nature-study is being taught; putting the word *dandelion* into problems, where the words *stone, book, box* or *knife* might just as well be used, is only verbal substitution and will have little effect on the pupil's relation to dandelions except to make him dislike them.

Having known one kind of common plant, the pupil may well study plant societies—how plants live together, and why. Every distinct or separate area has its own plant society. There is one association for the hard-tramped door-yard—knotweed and broad-leaved plantain with interspersed grass and dandelions; one for the fence-row—briers and choke-cherries and hiding weeds; one for the dry open field—wire-grass and mullein and scattered docks; one for the slattern roadside—sweet clover and ragweed and burdock; one for the meadow swale—smartweed and pitchforks; one for the barnyard—rank pigweed and sprawling barn-grass; one for the dripping rock-cliff—delicate bluebells and hanging ferns and grasses. These categories may be indefinitely extended. We all know the plant societies, but we have not thought of them.

In every plant society there is one dominant note: it is the individuality of one kind of plant that grows most abundantly or overtops the others. Certain plant-forms come to mind when one thinks of willows, others when he thinks of an apple orchard, still others when he thinks of a beech forest. The farmer may associate "pussly" with cabbages and beets, but not with wheat and oats. He associates cockle with wheat, but not with oats or corn. We all associate dandelions with grassy areas, but not with burdocks or forests.

It is impossible to open one's eyes out-of-doors outside the paved streets of cities without seeing a plant society. A lawn is a plant society. It may contain only grass, or it may contain weeds hidden away in the sward. What weeds remain in the lawn? Only those that can withstand the mowing. What are they? Let a bit of lawn grow as it will for a month and see what there is in it. A swale, a dry hillside, a forest of maple, a forest of oak, a forest of hemlock or pine, a weedy yard, a tangled fence-row, a brook-side, a deep quiet swamp, a lake shore, a railroad, a river bank, a meadow, a pasture, a dusty roadway—each has its characteristic plants. Even in the winter one may find these societies—the tall plants still asserting themselves, others of less aspiring stature, and others snuggling just under the snow.

Later, special attributes or forms of plants may be considered—forms of stems, bark, ways of branching, root forms, leaf forms, position and size of leaves with reference to light, flower forms, falling of the leaves, germination, seed dispersal, pollination (for older pupils), injuries of various kinds (as by snow, ice, wind, sun-scalding, drought, insects, fungi, browsing by cattle), simple physiological experiments of many kinds (such as are now described in our best text-books). In winter, studies may be made of the forms of trees and bushes and of persisting weeds, leaf-buds and fruit-buds, bark forms, preparation for spring, tubers and bulbs, seed-sowing and germination, struggle for existence in the tree-top, evergreens and how they shed their leaves, how the different kinds of trees hold the snow, where the herbs and tender things are, cones and seed pods, apples and turnips and other things from the cellar, knots and knot-holes, how vines hold to their supports, and others. These subjects are intended only as suggestions of the kind of work that may be taken up with profit.[63]

As far as possible, the study of form and function should go together. Correlate what a part is with what it does. What is this part? What is its office, or how did it come to be? It were a pity to teach phyllotaxy without teaching light-relation: it were an equal pity to teach light-relation without teaching phyllotaxy.

There are those who discourage the teaching of plant societies until the pupil is well grounded in "physiology"; but this, again, is the science-teaching point of view. Of course the child cannot understand the fundamental reasons for plant association—I wonder whether the botanist does?—but the child can comprehend the phenomena, and he will be interested in them because they are so intimately associated with him and are observable.

There are those, again, who say that such subjects as those suggested above do not prepare the pupil to enter college. My reply is that the elementary schools do not exist for the sake of the college or the university. Those that are to enter college are a small and special class, and they may receive special instruction.

I have spoken of the herbarium stage of plant-study and have said that it is passing away. It is perfectly possible, however, to make herbaria without in any way lessening the value of beginning plant-work (the rather increasing its value), but the herbarium should be a result of the work rather than constitute the work itself. After the pupil has come to know the dandelion or a plant society or the flora of the neighborhood, he will do well to make specimens; these specimens will be a part of his records.

VI

The Growing of Plants by Children—The School-Garden

ACTUALLY to grow a plant is to come into intimate contact with a specific bit of nature.[64] The numbers of plants that we grow, and also the kinds of them, increase with every generation. The intensity of our plant-growing, as well as the increasing care for animals, is coming to be a measure of our interest in the world about us.

Not only has the cultivation of plants itself increased our contact with plants and with nature, but, in connection with the growth of the spirit of art, of sport, and of suburbanism, it has taken us afield and has impelled us to know things as they are and as they grow.[65] The modern popularization of plant-knowledge is probably due more to these agencies than to the progress of botany.

There are many practical applications to the lives of children and to the home that may be made from a knowledge of plants and horticulture. This knowledge means more than mere information of plants themselves. It takes one into the open air. It enlarges his horizon. It brings him into contact with living things. It increases his hold on life. All these facts were well understood by Froebel, Pestalozzi, and other educational reformers.[66]

It is important that one does not assume too much when beginning plant-work with children. We forget that things which fail to appeal to us, because of our busy lives and great experience, may nevertheless mean very much to the child. Often we attempt to teach the child so much that it is confused and nothing makes an impression. An interest in one simple, living problem that is near to the child's life is worth a whole book of facts about nature.

It is not primarily important that children know the names, although the name is an introduction to a plant as it is to a person. The essential point is that there should be plants about the home, or in the school grounds, or in the schoolhouse windows. Even though the children are not conscious that they are receiving any impression from these plants, nevertheless the very presence of them has an influence that will be felt in later life, even as the presence of good literature and furniture and the association of refined surroundings has influence.

> I dropped a seed into the earth. It grew, and the plant was mine.
>
> It was a wonderful thing, this plant of mine. I did not know its name, and the plant did not bloom. All I know is that I planted something apparently as lifeless as a grain of sand and there came forth a green and living thing unlike the seed, unlike the soil in which it stood, unlike the air into which it grew. No one could tell me why it grew, nor how. It had secrets all its own, secrets that baffle the wisest men; yet this plant was my friend. It faded when I withheld the light, it wilted when I neglected to give it water, it flourished when I supplied its simple needs. One week I went away on a vacation, and when I returned the plant was dead; and I missed it.
>
> Although my little plant had died so soon, it had taught me a lesson; and the lesson is that it is worth while to have a plant.[67]

Provide some little means of growing plants, not only to teach how to grow plants themselves, but to instruct the child in the care of things, to show that other beings besides itself have vicissitudes and lives of their own, and to implant the germ of altruism—the interest in something outside of oneself. These means of growing plants should be simple. A pot, a box or a hotbed may be sufficient. Every child should have the handling of at least one plant during the period of childhood. One plant cannot be handled without leaving an impression on the life.

The love of plants should be inculcated in the school. It can usually be better done in school than at home, particularly when one or both of the parents is opposed to it and constantly discourages the child. Even when the parents are ready and competent, the teacher may be able to reach the children more effectively than they. In nearly every school it is possible to have a few plants in the window. They may not thrive, but it is worth while to set the children to inquiring why they do not. Sometimes the poorest plants awaken the most effort and inquiry. If nothing else will thrive, a beet will. Secure a good fresh beet-root from the cellar. Plant it in a box or tin can. Surprisingly quick it will throw out clean bright leaves. The thick root will hold moisture from Friday to Monday.

A desire for school-gardens is gradually taking shape. This movement must grow and ripen; it cannot be perfected in a day. Through the centuries there have been few school-gardens: we must not expect to overcome the lack at once. The movement has not been aided much, if at all, by those who have "complete" schemes for gardens for the district schools. Such schemes may be advisable later. Start the work by suggesting that the school-grounds be cleaned or "slicked up." Take one step at a time. The propaganda for school-gardens must have relation to the economic and social conditions under which the school exists.

There is some confusion as to the objects of school-ground improvement. The purposes may be analyzed as follows:

(1) Ornamenting the grounds, comprising (*a*) cleaning and tidying them, (*b*) securing a lawn, (*c*) planting. This is always the first thing to be done. It stands for thrift, cleanliness, comfort, beauty, progressiveness.

(2) Establishing a collection to supply material for nature-study and class work.

(3) Making a garden for the purpose of (*a*) supplying material (as in No. 2), (*b*) affording manual-training, object lesson work, and instruction in plant-growing.[68]

(4) Providing a test ground or experiment garden where new varieties may be tried, fertilizer and spraying experiments conducted, and other definite studies undertaken.

These purposes fall into two main groups: (1) The improvement or adornment of the grounds; (2) the making of distinct gardens for purposes of direct instruction, or school-gardening proper. Much of the current discussion does not distinguish these two ideals, and thereby arises some of the loss of effort and effectiveness in the movement.

Improvement of the school-grounds

Every school-ground should be picked up, cleaned up and made fit for children to see. There are three stages in the improving of any ground: Cleaning up; grading and seeding; planting.

To improve the school-grounds should be a matter of neighborhood pride. It is an expression of the people's interest in the things that are the people's. We are ashamed when our homes are not fit and attractive for children to live in; but who cares if at the school the fence is tumble-down, the wood or coal scattered over the yard, the clapboards loose, the chimneys awry, the trees broken, the outhouses sagged and yawning?

The first thing to do is to arouse the public conscience. Begin with the children. As soon as they are directed to see the conditions they will believe what they see. They are not prejudiced. They will talk about it: teacher, mother, father will hear.

The next step is to "clean up." Do not begin with any ideal plan of landscape-gardening improvement to be carried out at once—not unless some one person is willing to do all the work and bear all the expense out of his public spirit; and this would be unfortunate, because most of the value in improving a ground is to interest the children in the work. Develop the children's enthusiasm—it is easy to do—in removing stones and litter and rubbish, in filling the holes, piling the wood, raking the grounds. If one school year were required to accomplish this work alone it would be time well used. Children and teachers have many interests. We are likely to expect too much of them.

The cleaning up once done, and the civic pride aroused to the pitch of keeping it done, the next step is to make a base or foundation upon which all the gardening or planting features are to stand: the land must be graded. In some cases the soil must be removed and new earth put in its place, for the soil about a schoolhouse is very likely to be poor sand or clay, or a mixture with building material and other rubbish; but in general this labor will not be necessary if only a lawn and ornamental planting are desired. In some places a lawn is impracticable, but a good and even earth surface should always be secured. The early spring is the season in which to do all this shaping and seeding of the land. The spring fever is on and enthusiasm is new-born. If the school is in the country, the farmers can be interested to do the heavy work. If the subject has been well discussed in the school for some weeks or months, it should not be difficult to organize the farmers into a "bee" to grade, till and seed the ground. There is always at least one energetic man in the community who is ready to take the lead in such movements as this. Much of the value of improving the school-ground lies in its arousing of public interest.

The next year, plant. Let the matter be discussed in school. Ask the children to make plans. When the time is ready, choose the simplest plan that seems to fulfil the requirements. It is well to get expert advice on this plan. Remember that during a large part of the year the school-ground will be practically without care; the planting must be able to maintain itself, if necessary. Leave the centers open. Throw the planting mostly to the borders or margins. Be careful not to have scattered effects in planting. Have the planting as little and as simple as possible and yet accomplish the desired results. Avoid all elaborate designs in bedding. Leave ample space for playgrounds. Cover the out-buildings with vines, and screen them with bushes and trees. Use chiefly of hardy and well-known trees and shrubs and herbs. Aim to have the ground interesting because it appeals to the onlooker as a picture and not as a collection of plants.[69]

The school-garden

The real school-garden is for direct instruction. It is an outdoor labora-tory. It is a part of the school equipment, as books, blackboards, charts and apparatus are. The school-garden is not adapted to all schools; or, to speak more correctly, not all schools are yet adapted to the school-garden, any more than they are all ready for an equipment in physics or chemistry. All grounds can be improved and embellished; we shall be glad when all schools will also have a school-garden.[70] The making of a definite garden is an epoch in the life of each school: it marks the progress of the school in educational ideals.

The school-garden should have a special area set aside for it, as any other garden, room or laboratory has. Its prime motive is not to be ornamental, but to be useful.[71] The garden should be a good garden, if it is to do its best work.[72] By this I do not mean that it be perfect from the gardener's standpoint, but that it be carefully planned and the ground put in good condition. The children should do the gardening; a gardener or teacher should not take care of the children's beds for them. (For a description of actual school-garden work, see p. 177.)

A school-garden has a large range of usefulness. It supplants, or, at least, supplements mere book training; presents real problems, with many interacting influences, affording a base for the study of all nature, thereby developing the creative faculties and encouraging natural enthusiasm; puts the child into touch and sympathy with its own realm; develops manual dexterity; begets regard for labor; conduces to health; expands the moral instincts by making a truthful and intimate presentation of natural phenomena and affairs; trains in accuracy and directness of observation; stimulates the love of nature; appeals to the art-sense; kindles interest in ownership; teaches garden-craft; evolves civic pride; sometimes affords a means of earning money; brings teacher and pupil into closer personal touch; works against vandalism; aids discipline by allowing natural exuberance to work off; arouses spontaneous interest in the school on the part of both pupils and parents; sets ideals for the home, thereby establishing one more bond of connection between the school and the community.*

The larger relations

There is a broader significance to the growing of plants, as indicated in the foregoing catalogue, than that associated with mere garden-making or with the furnishing of schoolroom material alone. There are social and national aspects. Children in the home and school should be interested in horticulture and agriculture as a means of introduction to nature. Farming introduces the human element into nature and thereby makes it more vivid

---

* From "*Outlook to Nature*," p. 129 (Rev. Ed.)

in the child's mind. More than half the people of the United States live outside the cities. More persons are engaged in farming than in any other single occupation. The children in the schools are taught much about the cities, but little about the farming country. The child should be taught something from the farmer's point of view, and the teaching of gardening is one of the ways in which to begin. This will broaden the child's horizon and quicken his sympathies. Every person is now supposed to know something of the country. He will spend part of his vacations therein. The more knowledge he has of farming methods the more these vacations will mean. It is not necessary, and perhaps not even important, that the child be taught these subjects with the purpose of making him a farmer, but rather as a means of education and of interest to him in the out-of-doors.[73]

There must be a greater interest in parks and public gardens. These institutions have now come to be a part of our civic life. They no longer need apology. We build parks in the same spirit that we build good streets and make sanitary improvements; but the park should be more than a mere display of gardening. It should have an intimate relation with the lives of the people.[74] All parks should be open to nature-study teachers, at least on certain days. There should also be children's days in the parks. In some places the park may grow specimens for the school. In large cities some of the common vegetables and farm crops may be grown in small areas at one side of the park. The tendency, perhaps, is to make our parks too exotic, and to give relatively too much attention to mere roads, statuary, and architecture.

The general appearance and attractiveness of the home can be greatly improved by simple gardening. The perfect garden, from the gardener's point of view, may not be the most useful or most decorative one. The garden should be so common and so easy to make as to become a part of the child-life.[75]

VII

Nature-Study Agriculture

THE nature-study idea is bound to have a fundamental influence in carrying a vital educational impulse to farmers.[76] The accustomed methods of education are less applicable to farmers than to any other people, and yet countrymen are nearly half our population. The greatest of the unsolved problems of education is how to reach the farmer. He must be reached on his own ground. The methods and the results must suit his needs. The ultimate test of good extension work will be its ability to reach into the remotest districts.[77]

We have failed to reach the farmer effectively because we still persist in employing old-time and academic methods. Historically, the common public school is a product of the university and college. "The greatest achievement of modern education," writes W. H. Payne, "is the gradation and correlation of schools, whereby the ladder of learning is let down from the university to secondary schools, and from these to the schools of the people."[78] This origin of "the schools of the people" from the university explains why it is that these schools are so unrelated to the life of the pupil, and so unreal; they are exotic and unnatural. If any man were to find himself in a country devoid of schools and were to be set the task of originating and organizing a school system, he would almost unconsciously introduce some subjects that would be related to the habits of the people and to the welfare of the community. Being freed from traditions, he would teach something of the plants and animals and fields and people and affairs.[79]

So long have we taught the text-book routine that we do not seem to think that there may be other and better means. We may allow the Greek idea of education for culture, but we must have other education along with it. It is possible to acquire culture at the same time that we acquire power.[80] Education for culture alone tends to isolate the individual; education for sympathy with one's environment tends to make the individual an integral part of the activities and progress of his time. At all events, there must be as great possibility for culture in the nature-studies as there is in the customary subjects of the common schools. My plea is that new educational methods must be employed before we can really reach the farming communities. I am not insisting that we make more farmers, but that we relate the rural school to the lives of people and that we cease to unmake farmers.

Man is a land animal and his connection with the earth, the soil, the plants, animals and atmosphere is intimate and fundamental. This earth-relationship is best expressed in agriculture,—not agriculture merely as a livelihood, but as the expression of the essential relationship of man to his planet home. Agriculture affords a primary educational course for the development of the race. If this kind of instruction is really to come and to be effective, nature-study agriculture is not to be added to the school work so much as to grow out of it as a redirection or reconstruction of it. The best agriculture is a perfect adaptation of man to his natural environment.

A point of view on the rural-school problem

A fundamental necessity to successful living is to be in sympathy with the nature-environment in which one is placed. This sympathy is born of good knowledge of the objects and phenomena in the environment. The process

of acquiring this knowledge and of arriving at this sympathy is now popularly called nature-study.

The nature-study process and point of view should be a part of the
work of all schools, because schools train persons to live. Particularly
should it be a part of rural schools, because the nature-environment is
the controlling condition for all persons who live on the land. There is
no effective living in the open country unless the mind is sensitive to the
objects and phenomena of the open country; and no thoroughly good
farming is possible without this same knowledge and outlook. Good
farmers are good naturalists.

Inasmuch as this nature-sympathy is fundamental to all good farming,
the first duty of any movement is to establish an intelligent interest in the
whole environment,—in fields and weather, trees, birds, fish, frogs, soils,
domestic animals. It would be incorrect to begin first with the specific
agricultural phases of the environment, for the agricultural phase (as any
other special phase) needs a foundation and a base: it is only one part of a
point of view. Moreover, to begin with a discussion of the so-called "useful" or "practical" objects, as many advise, would be to teach falsely, for,
as these objects are only part of the environment, to single them out and
neglect the other subjects would result in a partial and untrue outlook to
nature; in fact, it is just this partial and prejudiced outlook that we need
to correct (p. 90).

The colleges of agriculture have spread the nature-study movement.
Such work was begun as early as 1895 and 1896 by the College of Agriculture of Cornell University.[81] The colleges would have been glad if
there had been sufficient nature-study sentiment to have enabled them
to emphasize the purely agricultural phases in the schools; but this sentiment had to be created or quickened. At first it was impossible to secure
much hearing for the agricultural subjects. Year by year such hearing has
been more readily given, and the work has been turned in this direction as
rapidly as the conditions would admit,—for it is the special mission of an
agricultural college to extend the agricultural applications of nature-study.

In making these statements I have it in mind that the common schools
do not teach trades and professions. I would not approach the subject
primarily from an occupational point of view, but from the educational
and spiritual; that is, the man should know his work and his environment.
The mere giving of information about agricultural objects and practices

can have very little good result with children. The spirit is worth more than the letter. Some of the hard and dry tracts on farming would only add one more task to the teacher and the pupil if they were introduced to the school, making the new subject in time as distasteful as arithmetic and grammar often are. In this new agricultural work we need to be exceedingly careful that we do not go too far, and that we do not lose our sense of relationships and values. Introducing the word agriculture into the scheme of studies means very little; what is taught, and particularly how it is taught, is of the greatest moment. I hope that no country-life teaching will be so narrow as to put only technical farm subjects before the pupil.

We need also to be careful not to introduce subjects merely because practical grown-up farmers think that the subjects are useful and therefore should be taught. Farming is one thing and teaching is another. What appeals to the man may not appeal to the child. What is most useful to the man may or may not be most useful in training the mind of a pupil in school. The teacher, as well as the farmer, must always be consulted in respect to the content and the method of teaching agricultural subjects. We must always be alert to see that the work has living interest to the pupil, rather than to grown-ups, and to be on guard that it does not become lifeless. Probably the greatest mistake that any teacher makes is in supposing that what is interesting to him is therefore interesting to his pupils.

It has recently been said that the nature-study idea must disappear in rural schools and that agriculture must take its place. Nothing can be farther from the mark. Nature-study may be directed more strongly in agricultural applications, as the schools are ready for it, but the process is still nature-study. All good agricultural work in the grades must be nature-study.

All agricultural subjects must be taught by the nature-study method, which is: to see accurately; to reason correctly from what is seen; to establish a bond of sympathy with the object or phenomenon that is studied. One cannot see accurately unless one has the object itself. If the pupil studies corn, he should have corn in his hands and he should make his own observations and draw his own conclusions; if he studies cows, he should make his observations on cows and not on what some one has said about cows. So far as possible, all nature-study work should be conducted in the open, where the objects are. If specimens are needed, let the pupils collect them. See that observations are made on the crops in the field as well as on the specimens. Nature-study is an out-door process: the school-room should be merely an adjunct to the out-of-doors, rather than the out-of-doors an adjunct to the schoolroom, as it is at present (pp. 94, 101, 108).

A laboratory of living things is a necessary part of the best nature-study work. It is customary to call this laboratory a school-garden. We need to distinguish different types of garden (page 114): (1) The ornamental or planted grounds; this should be a part of every school enterprise, for the premises should be attractive to pupils and they should stand as an example in the community. (2) The formal plat-garden, in which a variety of plants is grown and the pupils are taught the usual handicraft; this is the prevailing kind of school-gardening. (3) The problem-garden, in which certain specific questions are to be studied, in much the spirit that problems are studied in the indoor laboratories; these are little known at present, but their number will increase as school-work develops in efficiency; in rural districts, for example, such direct problems as the rust of beans, the blight of potatoes, the testing of varieties of oats, the study of species of grasses, the observation of effect of fertilizers, may well be undertaken when conditions are favorable, and it will matter very little whether the area has the ordinary "garden" appearance. In time, ample grounds will be as much a part of a school as the buildings or seats now are. Some of the school-gardening work may be done at the homes of the pupils, and in many cases this is the only kind that is now possible; but the farther removed the laboratory, the less direct the teaching.

To introduce agriculture into any elementary rural school, it is first necessary to have a willing teacher. The trustees should be able to settle this point. The second step is to begin to study the commonest and most available object concerning which the teacher has any kind of knowledge. The third step is to begin to connect or organize these observations into a plan or system. This simple beginning made, the work ought to grow. It may or may not be necessary to organize a special class in agriculture; the geography, arithmetic, reading, manual-training, nature-study and other work may be modified or re-directed. It is possible to teach the state elementary syllabus in such a way as to give a good agricultural training.

In the high-school, the teacher should be well trained in some special line of science; and if he has had a course in a college of agriculture he should be much better adapted to the work. Here the teaching may partake more of the indoor laboratory method, although it is possible that our insistence on formal laboratory work in both schools and colleges has been carried too far. In the high-school, a separate and special class in agriculture would better be organized, and this means, of course, the giving up of something else by the pupil.

In many districts the sentiment for agricultural work in the schools will develop very slowly. Usually, however, there is one person in the community who is alive to the importance of these new questions. If this person has tact and persistence, he ought to be able to get something started. Here is an opportunity for the young farmer to exert influence and to develop leadership. He should not be impatient if results seem to come slowly. The work is new: it is best that it grow slowly and quietly and prove itself as it goes. Through the grange,[82] reading-club, fruit-growers' society, creamery association, or other organization the sentiment may be encouraged and formulated; a teacher may also be secured who is in sympathy with making the school a real expression of the affairs of the community; the school premises may be put in order and made effective; now and then the pupils may be taken to good farms and be given instruction by the farmer himself; good farmers may be called to the schoolhouse on occasion to explain how they raise potatoes or irrigate their land. A very small start will grow by accretion if the persons who are interested in it do not lose heart; and in five years every one will be astonished at the progress that has been made.

The prospect

In recent years there has been a marvelous application of knowledge and research to agricultural practice. We have exerted every effort to increase the productiveness and efficiency of the farm, and we have entered a new era in farming—a fact that will be more apparent in the years to come than it is now. The burden of the new agricultural teaching has been largely the augmentation of material wealth. Hand in hand with this new teaching, however, should go an awakening to the less tangible but equally powerful things of the spirit. More attractive and more comfortable farm homes, better reading, more responsive interest in the welfare of the community and the events of the world, closer touch with the common objects about him—these must be looked to before agriculture really can be revived. Appeal to greater efficiency of the farm alone cannot permanently relieve the agricultural status. This is all well illustrated in the attitude of children toward the farm. In a certain rural school in New York state of say forty-five pupils, I asked all those children that lived on farms to raise their hands: all hands but one went up. I then asked all those who wanted to live on the farm to raise their hands: only that one hand went up. Now, these children were too young to feel the appeal of more bushels of potatoes or more pounds of wool, yet they had this early formed their dislike of the farm. Some of this dislike is probably only an ill-defined desire for a mere change, such as one finds in all occupations, but I am convinced that the larger part of it was a genuine dissatisfaction with farm life. These children felt that their lot was less attractive than that of other children; I concluded that a flower-garden and a pleasant yard would do more to content them with living on the farm than ten more bushels of wheat to the acre. Of course, it is the greater and better yield that will enable the farmer to supply these amenities; but at the same time it must be remembered that the increased yield does not itself awaken a desire for them. I should make farm life interesting before I make it profitable.

It will be seen at once that all these new ideals are bound to result in a complete revolution or re-direction of our current methods of rural school-teaching. The time cannot be very far distant when we shall have systems of common schools that are based on the fundamental idea of serving the people in the very lives that the people are to lead. In many places there are strong protests against the old order; in other places there are distinct beginnings of the new order.

The beginnings of the new order are seen in the nature-study movement, the establishing of special agricultural schools, the strong agitation for county or district industrial schools, the spread of reading-courses, the rise of pupils' gardens, the extension work of the colleges of agriculture, the general awakening of rural communities. Books and methods are now derived for town schools rather than for country schools; the real texts for the rural schools are just now beginning to appear, and they represent a new type of school literature. In the future, the text-book is to have relatively less influence than in the past. We have been living in a text-book and museum age. All this old method is not to be complained of. The fact that so many new subjects and propaganda are coming in shows that we are in the midst of an evolution: we are in the making of progress.

Nature-study teaching may seem to be an indirect way of reaching the farmer; but it is not. It is direct because it strikes at the very root of the difficulty. Nature-study teaches the importance of actually seeing the thing and then of trying to understand it. The person who really knows a pussy-willow will know how to become acquainted with a potato-bug. He will introduce himself. One of the most significant comments I have heard on nature-study work came from a country teacher who said that because she had taught it, her pupils were no longer ashamed of being farmers' children. If only that much can be accomplished for each country child, the result will be enough for one generation. What can be done for the country child can be done, in a different sphere, for the city child. Fifty years hence the result will be seen.

A nature-study movement alone is not sufficient to awaken and reconstruct the agricultural interests. There should be coördinate efforts outside the schools. It particularly devolves on the colleges of agriculture to develop good extension teaching. The extension movement is already under way, several immediate causes combining to make it imperative, as (1) the people are ready for the work: they want to learn; (2) certain persons are ready to do the work: they want to teach; (3) the states appropriate money: the appropriations are made because work is done. Of these factors, the money is the least. No institution is so poor that something cannot be done if only the first three requisites are present. Time by time, perhaps little by little, the money will come. The work must be born, grow and mature.

This new teaching for the farmer is a most attractive field for well-directed effort. We need more teachers for it in the colleges and normal schools and common schools. The teaching in our agricultural colleges should be seized with the missionary spirit, with the desire to send out young persons who care not so much to make professors and experimenters in the great institutions, as to give themselves to spread the gospel of nature-love and of self-respecting, resourceful farming through all the colleges and all the public schools. The time is coming quickly when the college or school that wants really to reach the people must teach rural subjects from the human point of view.

We are on the borderland of a mighty country: we are waiting for a leader to take us into it.[83]

Part II

Containing several pieces that attempt to direct the teacher's outlook to nature

I

The Teacher's Interpretation of Nature

TWO sisters stood on the doorstep bidding good-by to their husbands, who were off for a day's outing.[84] One looked at the sky and said: "I am afraid it will rain." The other looked at the sky and said: "I know that you'll have a good time." There was one sky, but there were two women. There were two types of mind. There were two outlooks on the world. There are many persons who will not be pleased if they can help it.

I know a nature-study teacher whose first inquiry about any object is, "What is it worth?" Or, "What value has it to mankind?" Some objects are to be studied and protected because they are useful to man in supplying his wants, and all others are passed over as not worth knowing. I doubt whether this attitude can bring about any close and satisfying touch with nature. The long-continued habit of looking at the natural world with the eyes of self-interest—to determine whether plants and animals are "beneficial" or "injurious" to man—has developed a selfish attitude toward nature, and one that is untrue and unreal (pp. 90, 121). The average man to-day contemplates nature only as it relates to his own gain and enjoyment.

The satisfaction that we derive from the external world is determined by the attitude in which we consider it. All unconsciously one's habit of mind toward the nature-world is formed. We grow into our opinions and habits of thought without knowing why. It is therefore well to challenge these opinions now and then, to see that they contain the minimum of error and misdirection.[85]

The greatest thing in life is the point of view. It determines the current of our lives.

However competent a person may be in biology or other science, he cannot teach nature-study unless he has a wholesome personal outlook on the world.

The more perfect the machinery of our lives, the more artificial do they become. Teaching is ever more methodical and complex. The pupil is impressed with the vastness of knowledge and the importance of research. This is well; but at some point in the school-life there should be the opening of the understanding to the simple wisdom of the fields. One's happiness depends less on what he knows than on what he feels.

In these increasing complexities we need nothing so much as simplicity and repose. In city or country or on the sea, nature is the surrounding condition. It is the universal environment. Since we cannot escape this condition, it were better that we have no desire to escape. It were better that we know the things, small and great, which make up this environment, and that we live with them in harmony, for all things are of kin; then shall we love and be content. The growing passion for country life and the natural unspoiled world is a soul-movement.[86]

More and more, in this time of books and reviews, do we need to take care that we think our own thoughts. We need to read less and to think more. We need personal, original contact with objects and events. We need to be self-poised and self-reliant. The strong man entertains himself with his own thoughts. No person should rely solely on another person for his happiness.

The power that moves the world is the power of the teacher.

II

SCIENCE FOR SCIENCE'S SAKE

A DEMURE little woman at the teacher's convention told of the enthusiasm with which her pupils had collected butterflies and plants, and she described the museum that they had made.[87] She showed a folio of mounted
plants, and a cigar-box containing insects. I admired the specimens, and
mentally I complimented her judgment in finding so good use for such a
box. The tobacco odor kept the carnivorous bugs way, and I also com-
mended the judgment of the bugs. There was genuine enthusiasm in the
little woman's manner, and I wanted to be a young naturalist. When she
was talking, I strayed far in the fields and picked a dandelion.

But there was a man in the audience who squelched the little woman.
Her methods were all wrong. They were worse than wrong: the children
must unlearn what she had taught them. She should have begun with
some definite subject, and followed it systematically and logically. The
pupil must be held to the task day after day, until he masters the topic. To
skip from subject to subject is to be superficial. This way of teaching does
not result in mental drill. To make a collection is only play, and names are

vulgar. The pupil must be impressed with the completeness of his subject, and, above all things, he must be accurate. When he was talking, I smelled alcohol and I saw a frog in a museum jar.

Which was right? No doubt each was correct from the personal point of view, but wrong from the other's point of view. I recalled that the little woman recited only what she had done; the man upbraided her for not doing something else. Perhaps it is easy to advise and to criticize. The little woman was teaching children. She wanted to lead them to love the things they saw. She approached the subject from the human side, for are not the boy and the girl a part of what we call nature? They are not yet tamed and conventionalized. Does not every boy and girl like to go in the fields and "get" things? She was not thinking of the subject-matter; or if she did think of it, she knew that it could take care of itself. All she was thinking of—poor soul!—was to interest and educate the children. And she knew that if she set a subject and followed it unremittingly day by day the seats would soon be vacant.

The man was thinking of his college students; perhaps he had not considered that these students already liked the subject and needed only instruction. He forgot that you cannot force a person to choose a thing, although you may force him to take it. His were picked students, one from this town and another from that; hers were all the pupils in her little community. His pupils had seen and had chosen; to hers the world was all unseen and untried. His were the one in a hundred; hers were the entire hundred. His students had elected the subject; for this subject perhaps they were to live; they would increase the boundaries of knowledge; they would be scientists. He did not consider that all pupils would not be scientists.

Sometimes it seems as if scientists assume that they have the right of way in the subjects which they espouse; but there is more than one way of interpreting nature.[88] This domination is well illustrated in the usurpation of common words. The word "organic" relates to organisms and their products. But when the chemist studies the composition of organic compounds he defines the word in terms of chemistry. To him an organic compound may be a carbon compound or a carbohydrate derivative; and he can make an organic compound without any relation to an organism! Organic is a biological, not a chemical idea. Again, our forefathers used the word "bug" for many kinds of insects; but scientists have taken this word "bug" and have made it mean only a particular kind of a bug. This is all well enough amongst themselves, but when they attempt to make all the rest of the world use "bug" as they do, they go too far. Our forefathers have prior claims. It would be better if newly-made words could be used for new ideas. Science needs a technical language of its own.

What is the kernel of all this discussion about the pedagogical sin of making collections and of attaching names? It is no doubt derived from the older practice of merely naming things. The old idea of the study of nature was to make an inventory of the objects in the world. The objects are bewilderingly numerous, and to put them away in a cabinet, with a proper ticket attached, was to know them. The great want was names and classification; and these names must be arranged in books. This natural history bookkeeping received its largest impetus from the binomial method of naming, which might be called a system of "double entry."

This naming of objects is necessary. It is the starting-point, as a city directory is. But it is only the beginning of wisdom. It is not an end. The speculations of the modern evolutionists have emphasized the importance of the objects themselves in a new way. The point of view has changed. Do not let your pupils make an herbarium, the modern teacher may say, but tell them to study the plants. We all sympathize with this point of view; but what are we going to do with this native and exuberant desire of the child to explore and to collect?[89] And what better way is there to know plants and animals than actually to collect and to study them? One of my friends will not let his little boy make an herbarium, because that is mere superficial amusement; so the child collects postage stamps. He does not care to have him know the names of plants, but he is very careful to have him properly introduced to visitors; and what is an introduction but a conventional passing of names (p. 173)?

I think that science teaching has gone too far in discouraging the making of collections. We can make the collecting the means of securing real information. We can fasten the attention of the child. The one caution is not to make it an end. The child cannot collect without seeing the object as it lives and grows. It appeals to him more in the field than it does in the museum. Let him collect for the purpose of understanding a problem. Where does the dandelion grow? What are the plants in the bog? How many are the weeds in the orchard? What are the borers in the old log? Set the child a field problem and he will collect in spite of himself. Teach him at the same time to respect the rights of every living thing, and never to be wanton. Then the collecting has teaching power. But to make a collection of one hundred specimens in order to obtain a pass-mark is scarcely worth the effort (p. 111).

The point I urge is that there is no reason in the nature of things why subjects always should be taught this way or that, so long as they are taught truthfully and with purpose—and there are many ways of teaching the truth.[90] At one time or place we may teach for science's sake; at another time or place with equal justification we may teach for the pupil's sake.

III

Extrinsic and Intrinsic Views of Nature

"THE purpose of this exercise is to tell children how to see the hidden beauties of flowers."[91] Thus ran the announcement at the opening of the classroom period. Is it worth while to tell them any such thing? Why not teach them to be interested in plants? Why give them a half-truth when they might have the whole truth?

The "beauty" of a flower or a bird is only an incident: the plant or the bird is the important thing to know. Beauty is not an end. The person who starts out to see beauty in plants is often in the condition of mind that the dear old lady was who came into my conservatory and exclaimed, as she saw the geraniums, "Oh, they are as pretty as artificial flowers!"

But these people are not looking for beauty, after all; they look for mere satisfying form or color or oddity. They confound beauty with prettiness or with outward attractiveness. Real beauty is deeper than sensation. It inheres in fitness of means to end as well as in striking features. The child should see the object itself before he sees its parts or its attributes. Teach first the whole bug, the whole bird, the whole plant, with something of the way in which it lives. The botanist may well devote his life to a cell, but the layman wants to know the trees and the woods.

I dislike to hear people say that they love flowers. They should love plants; then they have a deeper hold. Intellectual interest should go deeper than shape or color. Teachers or parents ask the child to see how "pretty" the object is; but in most cases the child wants to know how it lives and what it does.

It is instructive to note the increasing love for wild animals and plants as a country grows old and mature. This is particularly well illustrated in plants. In pioneer times there are too many plants. The effort is to get rid of them. The forest is razed and the roadsides are cleaned. The pioneer is satisfied with things in the gross. If he plants at all, he usually plants things exotic or strange to the neighborhood. The woman grows a geranium or fuchsia in a tin can, and now and then makes a flower-bed in the front yard; but the man is likely to think such things beneath him. If a man has flowers at all, he must have something that will fill the eye. Sunflowers are satisfying.

But the second and third generations begin to plant forests and to allow the roadsides to grow wild at intervals. Persons come to be satisfied with their common surroundings and to derive less pleasure from objects merely because they are unlike their surroundings. Choice plants come into the yards here and there, and the men of the household begin to care for them. The birds and wild animals are cherished. (I know a man who in his pioneer days took no interest in crows except to get rid of them, but who later in life wept when a crow's nest in an apple tree was robbed.) Love of books increases. All this marks the growth of the intellectual and spiritual life.

America is a land of cut flowers. Nowhere does the cut-flower trade assume such commanding importance. Churches and homes are decorated with them. One sees the churches of the Old World decorated with plants in pots or tubs. The Englishman or the German loves to care for the plant from the time it sprouts until it dies: it is a companion. The American snips off its head and puts it in his buttonhole: it is an ornament. I have sometimes wondered whether the average flower-buyer knows that flowers grow on plants.[92]

All of us have known persons who derive more satisfaction from a poor plant that never blooms than others do from a bunch of American Beauty roses at five dollars. There is individuality—I had almost said personality—in a growing, living plant, but there is little of it in a detached flower. And it does not matter so much if the plant is poor and weakly and scrawny. Do we not love poor and crippled and crooked people? A plant in the room on washday is worth more than a bunch of flowers on Sunday.

But the American taste is rapidly changing. Each year the florist's trade sees a proportionately greater demand for plants. The same change is seen in the parks and home grounds. Every summer more gross carpet-beds are relegated to those parts of the grounds that are devoted to curiosities, or they are omitted altogether, and in their stead are restful sward and attractive plant forms. Flowers are not to be despised, but they are accessories.

This habit of looking first at what we call the beauty of objects is closely associated with the old conceit that everything is made to please man: man is only demanding his own. It is true that everything is man's because he may use it or enjoy it, but not because it was designed and "made" for "him" in the beginning. This notion that all things were made for man's special pleasure is colossal self-assurance. It has none of the humility of the psalmist, who exclaimed, "What is man, that thou art mindful of him?"[93]

"What were these things made for, then?" asked my friend. Just for themselves! Each thing lives for itself and its kind, and to live is worth the effort of living for man or bug. But there are more homely reasons for believing that things were not made for man alone. There was logic in the farmer's retort to the good man who told him that roses were made to make man happy. "No, they wa'n't," said the farmer, "or they wouldn't 'a' had prickers." A teacher asked me what snakes are "good for." Of course, there is but one answer: they are good to be snakes.

Being human, we interpret nature in human terms. Much of our interpretation of nature is only an interpretation of ourselves. Because a condition or a motive obtains in human affairs, we assume that it obtains everywhere. The only point of view is our own point of view. Of necessity, we assume a starting-point; therefrom we evolve an hypothesis which may be either truth or fallacy. Asa Gray combated Agassiz's hypothesis that species were originally created where we now find them and in approximately the same numbers by invoking Maupertuis's "principle of least action"—"that it is inconsistent with our idea of divine wisdom that the Creator should use more power than was necessary to accomplish a given end."[94] The result may be secured with a less expenditure of energy than Agassiz's method would entail. But who knows that "our idea of the divine wisdom" is correct? It is only a human metaphor; but, being human, it may be useful.

Much of our thinking about nature is only the working out of propositions in logic, and logic is sometimes, I fear, but a clever substitute for truth. It is impossible to put ourselves in nature's place—if I may be allowed the phrase; that is, difficult to work from the standpoint of the organism that we are studying. If it were possible to get that point of view, it would be an end to much of our speculation; we should then deal with things as they are.

We hope that we are coming nearer to an intrinsic view of animals and plants; yet we are still so intent on discovering what ought to be, that we forget to accept what is.

IV

MUST A "USE" BE FOUND FOR EVERYTHING?

EVERY pupil had a plant of the spring buttercup.[95] The teacher called attention to the long fibrous roots, the parted leaves, the yellow flowers; but these parts were apparently only incidentals, for she touched them lightly. But the hairs on the stem and leaves were important. They must be of some use to the plant. What is it? Evidently to protect the plant from cold, for does not the plant throw up its tiny stem in the very teeth of winter? It was clear enough; and thus are we taught that not the least thing is made in vain. Everything has its place and use; it is our business to determine what the uses are.[96]

I wondered how these children would look on the plants and animals they meet, and what the great round world would mean to them. The blackberry has thorns to keep away the animals that would harm it; the rabbit has soft short fur that it may pass through brush and briers; the mud-turtle is flat so that it will not sink in the mud; the poison sumac has venom to protect it from those who would destroy it; the crow is black that it may not be seen at night; the nettle has stings to punish its enemies;

the dog fennel has rank scent to protect it from the browsing animals; certain insects have a zigzag flight to enable them to elude their enemies. All the world is as perfect as a museum!

I wondered what would happen if some inquisitive child were to ask what becomes of all the plants that have no thorns or hairs or poison or ill scent. What if he should ask why the thornless blackberry does not perish, or why the sumacs that are not poisonous still live, or if he should suggest that the dandelion comes up earlier in the spring than the buttercup and yet has no hairs on its soft flower-stem? As I wondered, a little hand went up. The teacher granted a question. "Pigweeds ain't got prickers," said the boy. I saw that the boy was a philosopher. "True enough," replied the teacher promptly, "but I am sure that it has something with which to protect itself."

Thereby I knew her point of view: she had made up her mind what to see, and it was necessary only to hunt until she saw it; and in this respect she was like many another. Persons seem to interpret the struggle for existence as a fight. It is a sanguinary combat between adults. Everything must protect itself with armor. A botanist, in writing a description of a new and strange plant, noted the peculiar spines and then remarked: "That these are of some use to the plant can hardly be doubted. Perhaps they serve to prevent the access of undesirable insects."

Nothing is easier than to find an explanation for anything; the only difficulty is to determine whether the explanation is true. I have just read in an old book that the reason why a particular kind of graft failed to grow was because of the "disappointment of the sap."[97] I laughed at the expression; and yet is it not as scientific as to say that the hairs exist to keep the crowfoot warm or that the sumac has poison to protect it from its enemies? The teacher may as well have said that Jimmie Brown has freckles so that the sun will not tan his skin; and the statement would be hard to disprove.

A teacher asked me whether it is not true that her cactus has spines in order to protect it from browsing animals. I told her that I did not know. As I was a stranger to her, she wondered at my ignorance. She wanted to know why I did not know. I told her that I had no good evidence that an animal ever wanted to browse on her cactus or its ancestors. Perhaps the cactus spines are older than the browsing animals. Perhaps there was some special condition or reason in geologic time. Perhaps the spines were in some way an incidental result of the contraction of the plant body, which contraction was associated with the necessity of reducing the evaporating surface in an arid climate. Perhaps a hundred things. She was surprised that I had to go into geologic time to bury my ignorance. She wanted cause and effect side by side, and in the present. Then she could see them. It is a bother to look behind for causes.

This is a typical case. This attitude toward nature comes almost daily to the teacher; in fact, it sometimes comes from the teacher. The mischief is increased by many popular books on science, and some of these books have been written by persons who have done noble work for truth.

This is one of the greatest faults with the popular outlook on nature— the belief that every feature of plant or animal has a distinct use in the present time and that one has only to look to be able to see what this use is. Persons often look at the little things and miss the big ones. They look for the hairs and miss the plant. They see the unusual and overlook the common.[98]

Having seen a feature of which the function is not evident, they assume a condition and jump at a conclusion. A plant has poison; various creatures eat plants; the creatures are killed by poison: therefore the plant has poison to protect itself from the creatures. Now, it may even be true that the poison does protect the plant, but there is no proof thereby that the poison was produced for that purpose. The physiologist may find that the poison in the given case is merely a waste product of some chemical metabolism, and that the plant is fortunate in getting rid of it. If the plant is now and then protected, the result is an incident. If it should appear that one kind of plant, by natural selection or otherwise, has developed poison in order to protect itself, the fact would be spread abroad in book and magazine, but it would not be stated that it was one case out of a thousand. The exception is enlarged into the rule.

Persons like to believe in perfect adaptation of means to ends, without a slip or break in the process. They assume that all organisms have definite protectional features. A teacher brought a flower and asked what mechanism it had to insure cross-pollination. I told her that I was not aware that it had any; and she was surprised. She asked what mimicry protection a certain animal had; I was obliged to make a similar reply. I wish that somebody would write a book about non-adaptations and misfits in nature.

No one knows what spines and thorns are "for," and the true naturalist does not ask the question. He does not assume that because they would protect a man they would also protect another animal or a plant. He wants to know how they came to be, and what is their significance in the development of this particular race. He wants proof that adaptations are adaptations. He sets to work to find out. He cannot find out as he rides by on his horse—especially if he rides a hobby-horse.

This everything-has-a-use dogma is in part a reaction from the teachings of Darwin and his followers.[99] The dogma of special creation was overthrown. We were told that organisms and attributes have persisted because of natural selection—because they were best fitted to persist. The result, in many cases, is perfect adaptation of every organ and attribute. There followed a special literature on adaptation, mimicry and the like. The precision and design of the special-creation theory was transferred to the adaptation theory. The examples may all have been true, but one result has been to lead persons to look for adaptations and mimicry everywhere, and to assume that they exist. What does it matter if there is no special creation?—there is complete and universal adaptation, vindicating the wisdom of the Creator and our notions of what ought to be are verified.

But some one will say, if there is natural selection and survival of the fittest, adaptation must follow as a consequence. Yes; but it does not follow that every part or feature of the organism is specially adapted, at least not at the present epoch of time. A strong feature may carry other features that are merely innocuous or even harmful, as a horse carries a rider; and then, if unfit features tend to pass away, these features are misfits and remnants until they have disappeared.[100]

V

THE NEW HUNTING

THE world is full of animals and plants.[101] Every animal and plant has the power to multiply itself many fold. Every one contends for an opportunity to live.

This contention forces the individual to live for itself. Self-preservation, it is said, is the first law of nature. The animal appropriates food, usurps territory, kills and even devours its contestants. It kills because it must. It is goaded by the whip of necessity. To live is the highest desire that it knows. Its acts need no justification.

Man also is an animal. He has come up from the world-fauna. On his way he contended hand to hand with the other animal creation. He killed from necessity of securing food. As he rose above his contestants, this necessity became less urgent. He has now obtained dominion, but he is not yet fully emancipated from the necessity of taking life. Perhaps complete emancipation will come.

The old desire to kill—first born of necessity—still lingers with men. We still have much of the savage in us. But now we kill also for "sport." Practically a new motive has been born into the world with man—the desire to kill for the sake of killing. One generation of white men is sufficient practically to exterminate the bison and several other species. All this needs justification. The lower creation is not the plaything of man.

We are still obliged to kill for our necessities. We must secure food and raiment. More and more we are rearing the animals that we would take for food. We give them less dangerous lives. We protect them from the severities of the struggle for existence. We remove them from the necessities of protecting themselves from violence. We take our own. There is here little question of morals. We give that we may take; and we take because we must.

To kill for mere sport is a very different matter: it lies outside the realm of struggle for existence. Too often there is not even the justification of fair play. Usually the hunter exposes himself to no danger from the animal that he would kill. He takes no risks. He has the advantage of long-range weapons. There is no combat. Over on the lake shore every spring I see great cones of ice, built up by the action of the waves. Several stalwart men have skulked behind them and lie secure from observation. A little flock of birds, unsuspecting, unprotected, harming no man, obeying the laws of their kind, skims across the water. The guns discharge. The whole flock falls, the mangled birds struggling and crying, and tainting the water with their blood as they are carried away on the waves, perhaps to die on the shores. There is a shout of victory and a laugh of satisfaction. Surely, man is the king of beasts!

But there is another and fairer side. The lack of feeling for wounded animals is often thoughtlessness. The satisfaction in hunting is often the joy of skill in marksmanship, the pleasure of woodcraft, the enthusiasm of being in the open, the keen delight in discovering the haunts and ways of the nature-folk. Many a hunter finds more pleasure in all these things than in the game that he bags. The great majority of hunters are gentle and large-hearted men. They are the first to discourage mere wantonness and brutality. Under their hand, certain animals are likely to increase, because they eliminate the rapacious species. To the true sportsman, hunting is not synonymous with killing. It is primarily a means of enjoying the free world of the out-of-doors. The nature-spirit is growing, and there are many ways of knowing the fields and woods. The camera and spy-glass are competing with the trap and gun; and in time they ought to gain the mastery. It is no longer necessary to shoot a bird in order to know it.

I must not be understood as opposed to all hunting with the gun or the rod. Every man has a right to decide these questions for himself. I wish only to suggest that there are other ways of getting satisfaction from an expedition or a camping trip. There was a time when animals were known mostly in museums, or in books that suggested museums. We now know them in woods and fields where they live. We know what they do, as well as what they are. Making pictures from stuffed specimens will soon be a thing of the past. Read any book of natural history of fifty years ago; then read one of to-day. Note the road by which we have come: this may color your own attitude toward the nature-world.

A new literature has been born. It is written from the out-of-doors viewpoint, rather than from the study viewpoint. Man is not the only, nor even the chief, actor. Even the stories of animals of the old time do not have the flavor of this bright new literature. Not so very long ago animal stories were told for the purpose of carrying a moral—they were self-conscious. Now they are told because they are worth telling. The real moral is the interest in the animal and the way in which it contrives to live, not in some literary custom that tries to make an application to human conduct. No longer can one write a good nature-piece without intimate knowledge of the animal or plant in the wild, and until he has tried to put himself in its place. Perhaps the old school of literary effort is not losing ground; but it is certain that the new is gaining. The new literature is founded on first-hand knowledge, but it embraces all the human sympathies. It is the outcome of the study of objects and phenomena. The first product was scientific literature. The second is the lucid resourceful nature-writing of the present day. There are new standards of literary excellence.

The awakening interest in the nature-world is strongly reflected in the game laws—for these laws are only an imperfect expression of the growing desire to let everything live its own life. The recent revulsion of feeling against the shooting of trapped pigeons, as expressed in agitations before state legislatures, is an excellent example in point. It is gratifying that a prominent place in the discussions for good game laws is taken by sportsmen themselves. It is recognized that hunting for sport must be kept within bounds, and that it must rise above mere slaughter of defenseless animals.

Another expression of this growing sympathy is exhibited in the reservation of certain areas in which animals are to be unmolested. It is most significant that while many country regions are practically shot clean of animal life, sometimes even to songbirds, the parks and other public properties in cities often support this wild life in abundance. Usually it is easier to study squirrels and many kinds of birds in the city parks than in their native wilds. To this awakening interest in the preservation of animals is now added the desire to preserve the wild flowers and to protect scenery. The future will see the wild animals and plants safely ensconced in those areas that lie beyond the reach of cultivated fields; and these things will be the heritage of the people, not of the hunter, marksman, and collector alone.

This desire to protect and preserve our native animals is well expressed in President Roosevelt's reference to the subject when discussing the forest preserves in his first message to Congress: "Certain of the forest reserves should also be made preserves for the wild forest creatures. All of the reserves should be better protected from fires. Many of them need special protection because of the great injury done by live stock, above all by sheep. The increase in deer, elk and other animals in the Yellowstone Park shows what may be expected when other mountain forests are properly protected by law and properly guarded. Some of those areas have been so denuded of surface vegetation by overgrazing that the ground-breeding birds, including grouse and quail, and many mammals, including deer, have been exterminated or driven away. . . . In cases where natural conditions have been restored for a few years, vegetation has again carpeted the ground, birds and deer are coming back, and hundreds of persons, especially from the immediate neighborhood, come each summer to enjoy the privilege of camping. Some at least of the forest reserves should afford perpetual protection to the native fauna and flora, safe havens of refuge to our rapidly diminishing wild animals of the larger kinds, and free-camping grounds for the ever-increasing numbers of men and women who have learned to find rest, health and recreation in the splendid forests and flower-clad meadows of our mountains. The forest reserves should be set apart forever for the use and benefit of our people as a whole, and not sacrificed to the short-sighted greed of a few."[102]

The enlargement of our sympathies is also well reflected in the many societies that aim to lessen cruelty to animals. This movement is an outgrowth of the rapidly growing feeling of altruism—the interest in others—which, in the religious sphere, has ripened into the missionary spirit and into toleration. The prevention of cruelty to animals is of more consequence to man than to the animals. They suffer less than we. Perhaps the movement is in danger here and there of degenerating into mere sentimentalism and faddism; but, on the whole, it is sane and useful, because it measures our increasing sensitiveness.

Hunting to kill is not necessarily cruel. The best hunting is that which kills quickly. The poorest—for both the hunted and the hunter—is that which prolongs the struggle. The "gamey" fish is the one most liked by anglers. The "sport" of catching him depends on his desperate struggle for life; and this struggle is often prolonged that the excitement may be greater! Nature herself could be indicted for cruelty were not her practices dictated by inevitable conditions; but this fact does not release man, who acts largely as a free and moral agent. In nature, many animals meet violent or tragic deaths. The bird of passage that cannot keep up with its fellows is caught by the hawk or owl. The weaklings and stragglers are taken. Raise the curtain of night and behold the tragedies. Where are the graves of the unfit?

Man is not responsible for the tragedies of nature; but he is responsible for the tragedies that he himself inflicts.

The practices of any age are but the expressions of the needs and motives of that age. Much of the hunting is dictated by the desire of profits in money, and these profits often depend on fashion. Mere fashion has been the cause of the practical extermination of species of birds; but public opinion is finally aroused to check it.[103] The demand for furs is leading to similar results. Many species of animals perish before the continued progress of civilization, by means of which the native haunts are destroyed. We must protect that which we need to grow for our own use. It is inevitable that the animal creation, as a whole, shall recede as the earth is subdued to man. But too often this creation has fallen long before its time—fallen as a result of unnecessary killing, and of a desire of bloodthirstiness that is unworthy of us.

The foregoing remarks are meant to illustrate what I think to be an enlarging vision of our own place in the world. The point of view is shifting. The spiritual factors have increasingly more influence in shaping the course of our evolution. In time we shall probably be released entirely from the necessity of taking animal life to supply us with food. This will come as a result of our enlarging spiritual outlook rather than as a result of agitations concerned with questions of diet or with any mere propaganda. It is said that the conformation of man's teeth shows that a flesh diet is necessary, but this only indicates what our evolution has been, not what it will be or what is now a necessity for us. The further evolution will come slowly, but whatever it may be, we have reason to think that our points of contact with the nature-world will strengthen and multiply.

VI

THE POETIC INTERPRETATION
OF NATURE

MERRILY swinging on brier and weed,
    Near to the nest of his little dame,
Over the mountain-side or mead,
    Robert of Lincoln is telling his name:
        Bob-o'-link, bob-o'-link,
        Spink, spank, spink;
Snug and safe is that nest of ours,
Hidden among the summer flowers.
        Chee, chee, chee.[104]

Robert of Lincoln is gaily drest,
    Wearing a bright black wedding-coat;
White are his shoulders and white his crest.
    Hear him call in his merry note:
        Bob-o'-link, bob-o'-link,
        Spink, spank, spink;
Look what a nice new coat is mine,
Sure there was never a bird so fine.
        Chee, chee, chee.

Robert of Lincoln's Quaker wife,
    Pretty and quiet with plain brown wings,
Passing at home a patient life,
    Broods in the grass while her husband sings:
        Bob-o'-link, bob-o'-link,
        Spink, spank, spink;
Brood, kind creature; you need not fear
Thieves and robbers while I am here.
        Chee, chee, chee.

Modest and shy as a nun is she;
    One weak chirp is her only note.
Braggart and prince of braggarts is he,
    Pouring boasts from his little throat:
        Bob-o'-link, bob-o'-link,
        Spink, spank, spink;
Never was I afraid of man;
Catch me, cowardly knaves, if you can!
        Chee, chee, chee.

Six white eggs on a bed of hay,
    Flecked with purple, a pretty sight!
There as the mother sits all day,
    Robert is singing with all his might:
        Bob-o'-link, bob-o'-link,
        Spink, spank, spink;
Nice good wife, that never goes out,
Keeping house while I frolic about.
        Chee, chee, chee.

Soon as the little ones chip the shell,
    Six wide mouths are open for food;
Robert of Lincoln bestirs him well,
    Gathering seeds for the hungry brood.
        Bob-o'-link, bob-o'-link,
        Spink, spank, spink;
This new life is likely to be
Hard for a gay young fellow like me.
        Chee, chee, chee.

Robert of Lincoln at length is made
   Sober with work, and silent with care;
Off is his holiday garment laid,
    Half forgotten that merry air:
      Bob-o'-link, bob-o'-link,
      Spink, spank, spink;
Nobody knows but my mate and I
Where our nest and our nestlings lie.
       Chee, chee, chee.

Summer wanes; the children are grown;
Fun and frolic no more he knows;
Robert of Lincoln's a humdrum crone;
   Off he flies, and we sing as he goes:
      Bob-o'-link, bob-o'-link,
      Spink, spank, spink;
When you can pipe that merry old strain,
Robert of Lincoln, come back again.
       Chee, chee, chee.*

This was the exercise that the children were having as I visited the school on a June morning.[105] It was the new old song by which Bryant is remembered of the country boy and girl. The children had seen and studied the bobolink. They had heard the liquid rattle of his song. They had seen the nest in the grass. They had watched for the Quaker wife. They had seen the purple-flecked eggs. They knew that Robert of Lincoln would leave them. The poem touched their hearts.

    With enthusiasm I related the experience to my friend, the teacher of biology in a college. He doubted the value of such work. He saw only danger in it. Such teaching tends to looseness of ideas. It makes the mind

---

*   From Complete Works of William Cullen Bryant.
          Published by D. Appleton & Co.

[Probably *The Poetical Works of William Cullen Bryant*, vol. 2, New York: D. Appleton, 1883, 41–43. The title of the poem is "Robert of Lincoln," first published in *Putnam's Monthly*, June 1855. —Editor's note.]

discursive. It does not fix and fasten the attention on the subject-matter. It is unscientific. The child could learn poetry by the yard, he said, and yet not know how many toes the bobolink has, nor the shape and size of its wings. The pupil gains no comparative knowledge of bird with bird. The poem is untrue. The bobolink is not "drest": he has no clothes. He has no wife: he is mated, not wed.

I could only reply that the bobolink's toes have little relation to men's lives, however much they may have to bobolinks' lives; but the bobolink may mean much to men's lives. To a man studying ornithology—and I wish there were more—the toes are important; but I am seeking a fresh and firmer hold on life. I should rather know the song of the bobolink than to know all about the structure of the bird; of course, I should prefer to know both, if I could. To be sure, I should study the bobolink before I studied the poem; but I should want a real bobolink, not a stuffed specimen. If I were obliged to choose between lessons on stuffed bobolinks and the poem, I should take the poem: there is more bobolink in it.

I like Bryant's lyric because it catches so much of the life of a bobolink. A scientific description could tell the facts better, but only ornithologists read scientific descriptions. Yet I have always wished that the poet had told the whole story. After the breeding season is past, the birds gather in flocks in the rice-fields and reeds of the South and are then known as rice-birds and reed-birds. In great numbers they are slaughtered for the market, and thereby the bobolink does not become an abundant species in the North. May we not add:

Far in the South he gathers his clans,
    Nor thinks of the regions of ice;
Too early yet for housekeeping plans,
    He rev'ls and gluttons in fields of rice.
      Rice-bird, bob-o'-link,
      Spink, spank, spink;
Hunter is waiting under the bloom,
Robert of Lincoln falls to his doom.
      Chee, chee, chee.

Spring comes: swinging on brier and weed,
    Near to the nest of his little dame,

Over the mountain-side and mead,
   Another proud groom is telling his name:
      Bob-o'-link, bob-o'-link,
      Spink, spank, spink;
The meadow belongs to wife and me—
Life is as happy as life can be.
      Chee, chee, chee.

This is the age of fact, and we are glad of it. But it may be also an age of the imagination.[106] There need be no divorce of fact and fancy; they are only the poles of experience. What is called the scientific method is only imagination trained and set within bounds. Compared with the whole mass of scientific attainment, mere fact is but a minor part, after all. Facts are bridged by imagination. They are tied together by the thread of speculation and hypothesis. The very essence of science is to reason from the known to the unknown.

There can be no objection to the poetic interpretation of nature. It is essential only that the observation be correct and the inference reasonable, and that we allow it only at proper times. In teaching science we may confine ourselves to scientific formulas, but in teaching nature we may admit the spirit as well as the letter. If I were making a teacher's program for the study of nature, I should want to include a course in English poetry. With pupils, however, one must be careful to have the poem exactly appropriate to the subject and the occasion.

One may not make a list of poems that are always to be used by teachers of nature-study for specified topics. The choice of the poem should lie with the particular teacher or the pupils. These poems should be used sparingly, and not at all when the teacher himself does not have poetic feeling by means of which to interpret them. Better no poems whatever than to have manufactured and idle sentiment. The trouble with much of the sentiment is that it gives us a wrong point of view.

In our day of science, people seem to be afraid of figures of speech. The scientist forbids us to personify; and this is well. But this spirit may be carried so far as to forbid metaphor and to condemn parables. Speech cannot be literally accurate. Even astronomers say that the sun sets, but we know that it does not.[107] To say that a potato-plant works all the season in order to provide for its offspring the next year is said to give a wrong conception of the plant because it implies motive. But does this picture mislead any one? Everybody knows that a potato-plant has no brains. Everybody knows that the statement conveys a truth. If the phrase is not justifiable, then it is a question whether I may say that a potato has eyes. Much of the objection to statements of this kind is mere quibbling (pp. 103, 132).

But, on the other hand, all such allegories must be true in spirit and in their teaching value. Much of the current writing of plants and animals by which human motives are implied, is productive of harm; but we should distinguish between metaphor, or mere literary license, and an untrue point of view. The ultimate test is whether the reader is led to believe what is not true. An animal or a plant may be represented as telling its own story without misleading any one, even as a character in a novel may speak in the first person; we need not imply human motives or human points of view in these cases: there remain only the questions as to whether this is really good literary taste, and whether it is the most effective way to reach the audience for which it is intended. In general, a direct and lucid presentation, without circumlocution and invention, is to be preferred; and this direct method allows of the full expression of sentiment and the poetic impulse.[108]

I protest against that teaching of nature which runs into thin sentimentalism, which makes the "goody-goody" part of the work so prominent that it becomes the child's point of view, whether the writing is in prose or verse.[109]

The spirit of science lends itself well to song. The concrete is not unpoetic. If in this day we apostrophize and personify nature less, we have improved in the spirit and intimacy of our song. The point of view gradually has shifted from human interest in natural things to the things themselves. We need a free nature poetry that will give us confidence and a firm hold on life.

VII

AN OUTLOOK ON WINTER

IN the bottom of the valley is a brook that saunters between oozing banks.[110] It falls over stones and dips under fences. It marks an open place on the face of the earth, and the trees and soft herbs bend their branches into the sunlight. The hang-bird swings her nest over it. Mossy logs are crumbling into it. There are still pools where the minnows play. The brook runs away and away into the forest. As a boy I explored it but never found its source. It came somewhere from the Beyond and its name was Mystery.

The mystery of this brook was its changing moods. It had its own way of recording the passing of the weeks and months. I remember never to have seen it twice in the same mood, nor to have got the same lesson from it on two successive days; yet, with all its variety, it always left that same feeling of mystery and that same vague longing to follow to its source and to know the great world that I was sure must lie beyond. I felt that the brook was greater and wiser than I. It became my teacher. I wondered how it knew when March came, and why its round of life recurred so regularly with the returning seasons. I remember that I was anxious for the spring to come, that I might see it again. I longed for the earthy smell when the snow settled away and left bare brown margins along its banks. I watched for the suckers that came up from the river to spawn. I made a note when the first frog peeped. I waited for the unfolding spray to soften the bare trunks. I watched the greening of the banks and looked eagerly for the bluebird when I heard his curling note somewhere high in the air.

Yet, with all my familiarity with this brook, I did not know it in the winter. Its pathway up into the winter woods was as unexplored as the arctic regions. Somehow, it was not a brook in the winter time. It was merely a dreary waste, as cold and as forbidding as death. The winter was only a season of waiting, and spring was always late.

Many years have come and gone since then. My affection for the brook gave way to a study of plants and animals and stones. For years I was absorbed in phenomena. But now mere phenomena and materials have slipped into a secondary place, and the old boyhood slowly reasserts itself. I am sure that I know the brook the better because I know more about the things that live in its little world; yet that same mystery pervades it and there is that same longing for the things that lie beyond. I remember that in the old days I did not mind the rain and the sleet when visiting the brook. I was not conscious that they were not a part of the brook itself. It was only when I began to dress up that the rain annoyed me. I must make a proper appearance before the world. From that time the brook and I grew farther apart. We are coming together again now. It is no misdemeanor to get wet if you feel that you are not spoiling your clothing. One's happiness is largely a question of clothes.

But the brook is one degree the better now just because it remains a brook all winter. The winter is the best season of the four because there is more mystery in it. There is a new and strange spirit in the air. There are strange bird-calls in the depths of the still white woods. There are strange marks in the new-fallen snow. There are soft noises when the snow drops from the trees. There are grotesque figures on the old fence. There is the warm brown pathway of the brook still winding up between oozing banks. In the spring there are troops of flower-gatherers along the brook. In the summer there are fishers at the deep pools. In the fall there are nut-gatherers and aimless wanderers. In the winter the brook and I are alone. We know.

Most of us, I fear, look on winter with some feeling of dread and apprehension. It is to be endured. This feeling is partly due to the immense change that comes with the approach of winter. The trees are bare. The leaves are drifting into the fence-rows. The birds have flown. The deserted country roads stretch away into leaden skies. The lines of the landscape become hard and sharp. Gusty winds scurry over the fields. It is the turn of the year.

To many persons, however, the dread of winter, or the lack of enjoyment in it, is a question of weather. We speak of bad weather, as if weather ever could be bad. Weather is not a human institution, and it is not to be measured by human standards. There is strength and mighty uplift in the roaring winds that go roistering over the winter hills. The cold and the storm are a part of winter, as the warmth and the soft rain are a part of summer. Persons who find happiness in the out-of-doors only in what we call pleasant weather have not found the great joys of the open fields.

We speak of winter as bare, but this is only a contrast with summer. In the summer all things are familiar and close; the depths are covered. The view is restricted. We see things near by. In the winter things are uncovered. Old objects have new forms. There are new curves in the roadway through the forest. There are steeper undulations in the footpath. Even when the snow lies deep on the earth, the ground-line carries the eye into strange distances. You look far down into the heart of the woods. You feel the strength and resoluteness of the framework of the trees. You see the corners and angles of the rocks. You discover the trail that was lost in the summer. You look clear through the weedy tangle. You find new knotholes in the tree-trunks. You penetrate to the very depths. You analyze, and gain insight.

Many times in warm countries I have been told that the climate has transcendent merit because there is no winter. But to me this lack is its disadvantage. There are things to see, things to do, things to think about in the winter as in the spring. There is interest in the winter wayside, in the hibernating insects, in the few hardy birds, and the deserted nests, in the fret-work of the weeds against the snow, in the strong outlines of the trees, in the snow-shapes, in the cold deep sky. To many persons these strong alternations of the seasons emphasize and punctuate the life. They are the mountains and the valleys.[111] The winter is a part of the naturalist's year.

The lesson is that our interest in the out-of-doors should be a perennial current that overflows from a fountain that lies deep within us. This interest is colored and modified by every passing season, but fundamentally it is beyond time and place. Winter or no winter, it matters not: the fields lie beyond.

Part III

Comprising a budget of replies to many questions of school people

Inquiries and Answers

PRACTICAL problems confront the teacher. However well he may understand the theory and however fully he may agree with it, a new difficulty arises every time that he attempts to teach. A child will ask a question that a philosopher cannot answer; but on every question the teacher must have a point of view. I frequently speak to teachers on means of teaching nature-study. For the time they are pupils and they ask questions: I am obliged to take a point of view, and some of these opinions I have made note of at the time. Questions come in the mail. Some of these many inquiries and answers are here reprinted, not because they may be correct, but because they may be suggestive; and it will not matter if they repeat or expand some of the statements on the earlier pages.

How shall I know what subjects to choose?

Let the children choose the subject now and then. Let them collect the specimens.

But they may bring things of which the teacher knows nothing. So much the better! These are sometimes best for nature-study. They leave the largest interrogation point. From any subject the teacher can develop a fact. If he does not know the interpretation, say so: the pupils will be the more interested (p. 96). The teacher will not lose standing by the confession, if he is honest. Persons lose standing by pretending to know what they do not know and by being caught at it. The child is relieved to know that there is something yet to be discovered.[112]

In general, choose the subjects you are best prepared to teach and that best express or touch the conditions in which your pupils live. Whatever the subject, be careful to teach it simply and with the least apparent effort. Do not elaborate too much, or inject too much borrowed information. Always tie to the object or the materials. Do not teach zoölogy without animals, botany without plants, geography without knowing the earth, astronomy without stars, any more than you would teach grammar without language.

But if the child choose the material, the subject will lack continuity: what then?

Nature is not consecutive except in her periods. She puts things together in a mosaic. She has a brook and plants and toads and insects and the weather all together. Because we have put the plants in one book, the brooks in another, and the bugs in another, we have come to think that this divorce is the logical and necessary order.

If all the things mentioned above are taught, then the life of the brook will be the thread that ties them all together (p. 96). It is well to introduce the pupil to a wide range of material, in order to increase his points of contact with the world.

Then would you give no heed to continuity?

How much or how little the continuity will depend on the teacher and the circumstance. With children, the temptation is to have too much rather than too little continuity. First of all, we must develop the child's experience. The higher the grade, the more the topics may be correlated and coördinated. I doubt whether a closely graded nature-study is really nature-study at all. For children, I believe in that continuity and consecutiveness that relate the subject to its place and season. In April, correlate the work with the opening of the spring; in October, with the coming of winter. Compare the nature-study of June with that of May. Relate it to the farm work or other activities of the neighborhood. With living things, the cycle of the year is the fundamental continuity. Life-history is continuity. The procession of nature continues the work.

Should nature-study give way to "fundamental" work?

[Suggestions in reply to a foreign correspondent who asks whether we succeed in America in "getting good nature-study in one-teacher schools"; what attitude we take toward "the old-fashioned object-lesson work"; whether teachers are not in "great danger of forgetting that much of the most fundamental nature-study concerns dead matter, e.g., the simple chemical and physical changes that water and air undergo in relation to daily life."]

If nature-study is a way of teaching, then we ought not to expect ever to arrive at a complete agreement of opinion and practice. At the present time we are not even united on the fundamental educational questions involved, although we are gradually coming nearer to a consensus of opinion.

Many persons expect to find in the United States a great number of schools in which nature-study is taught, meaning by that to find separate classes set aside for this particular kind of work. In very many schools this will be found; but I suspect the greatest results in the end are to come when the nature-study mode or method runs through the teaching of all the accustomed subjects in the school, gradually reorganizing and revitalizing them (p. 80).

A school with one teacher can handle nature-study work as well as the school with twenty teachers if the teacher arrives at the nature-study way of teaching. I mean by this that the quality of the teaching may be good, quite independent of its quantity. Of course, we do not find a subject or a class under the name of nature-study in the one-teacher schools to any extent. What I mean by the nature-study spirit is to teach the things nearest at hand in a natural way and with the welfare of the child always in mind.

I am sure that it is perfectly possible to teach a child correctly and to put him into direct and sympathetic touch with the world he lives in by beginning with the biological and general phases of his environment even though he does not know the underlying chemical and physical processes and reasons. In fact, I am convinced that we must give up the idea that the child at first must know the so-called fundamental processes before he can know objects and phenomena. As a matter of fact, not one of us in the world, even the best of us, really knows the fundamental facts. We have merely gone a little further than some others have gone, but in the end everything is relative. If our first object is to develop the child and to train his capacities and sympathies, then it may not be necessary at all to begin with the underlying or internal reasons of things. These reasons will come out as the child grows and as his mind is able to grasp them.

I hope that we are rapidly passing through the epoch of mere object-teaching.[113] It has very narrow limitations as ordinarily taught, because it has had no vital relation to the child or to the life that he is to lead. Merely to study an object may or may not be of value in the training of the child. If that object has some relation to the life that the child is living so that it will be meaningful to him, it ought to have direct value in interesting him and in being made a means of drawing him out into larger growth.

From these remarks it will be seen that we need not "replace" some of the "fundamental work," as you phrase it, by nature-study. I would have all work, fundamental and otherwise (including "the simple chemical and physical changes that water and air undergo in relation to daily life"), taught in the nature-study spirit.

What is the proper pedagogical starting-point for nature-study?

[Reply to an inquiry from an officer in a normal college, who is urged to develop the nature-study in accordance with a pedagogical hypothesis. He is advised as follows, and he asks an opinion:

"The first advice is from the standpoint of the biologist, that the child repeats the history of the race and therefore should go to that place in history for material which will correspond with the stage through which the child is passing. The nature-study work would be based upon this idea and the history and literature chosen as nearly as possible from the stages through which the child may be passing at a given time.

"The other point of view makes the child's present environment the standpoint for getting everything, and the child with this as a basis looks back upon and studies the life through which the race has passed. The first point of view is really an application of the culture-epoch theory[114] in many ways except that some of our people wish to use nature-study as the starting point instead of literature and history."]

I do not consider myself competent to answer any questions on abstract theories of pedagogy. I did not come to my present work through that route. My educational outlook has developed personally and is founded essentially on the needs of the child, as I have been able to estimate those needs, without reference to pedagogical theory. I have heard discussions of the culture-epoch theory and other hypotheses of the psychology of education, but I am always obliged to come back to the simple fact that the child lives in a real environment and that this environment should be known to him and appreciated by him. I do not depreciate the value of the psychological theories, but I am not able properly to place the nature-study work with reference to them.

I should teach the child's world as he knows it, for the purpose of enabling him to know it better and to understand it. I should establish the child in his own life and anchor the school to the actual necessities of the community. From this starting point, I go backward or forward as the necessities of the case seem to demand, without any particular reference to the abstract psychology of the process. The child is not conscious of his place in the history of the race until he is told of it; and when he is told of it, it is a bit of extraneous and exotic information, the same as any other extrinsic information is. Of course, the child can be greatly interested in this fact, as he can be in any other fact or set of facts under the inspiration of a first-class teacher; but this of itself does not appeal to me as being sufficient reason for instituting a method. From the teacher's side, I doubt whether it is good practice to use the child as a means of working out an hypothesis. It is natural that every specialist should consider his subject to be the center of the circle.

I should begin with the common and apparent facts of our existence and conditions, or with the next-at-hand; beginning at home, I should pursue the exploration, and try to educate the child by the process.

How shall I make a start?

Persons hesitate, fearing that they will make a mistake. A teacher asked me the other day where he should begin with nature-work. He had been considering the matter for two or three years, he said, but did not know how to undertake it. I replied, Begin! Head end, tail end, in the middle— but Begin! There are two essential epochs in any enterprise—to begin, and to get done.

For the first lesson, choose the natural object that you know most about. Every teacher has sufficient knowledge of one subject to afford one good nature-study lesson. The second lesson will take care of itself.

If you are a principal, supervisor or other administrative officer and are thinking of starting off a movement in all the schools in a city or a com- missioner's district or in a county, first choose your teachers. Choose those that have enthusiasm and "good spirit" and that are not tied hand and foot to customary methods. Choose the fearless teachers—the ones that are anxious to arouse the pupils even though they do not do it by the book. Then give these teachers one good lesson yourself. Or, if you cannot give the lesson, put in their hands one good nature-study leaflet. Choose the leaflet as you would a teacher—for cheery outlook, energy, and directness of expression. Choose a leaflet that sends the teacher directly to nature; you do not want stories. Choose the leaflet that has snap and spirit, not mere information. It should be attractive in subject-matter and in mechan- ical execution. Never put a cheaply illustrated and poorly printed leaflet before a pupil. Remember that children are optimists, and that they want the best in both teacher and leaflet. Let the teacher study the object and the leaflet until the subject is mastered. When the teacher is full of the subject, he cannot help teaching.

If you are fortunate enough to have the starting of a nature-study movement for a State or other large territory, buy a small quantity of one of the best leaflets you can find. If you do not have the money, borrow it. Send a note to the newspapers to the effect that any teachers who wish to take up nature-study work may write you for literature and advice. All the rest will work itself out. Money will come from some source. Soon you will be publishing leaflets of your own; but be careful who writes them.

Beware of putting your trust in leaflets alone. Follow them up with cor- respondence and other personal work. The leaflet will not work of itself. It will soon be forgotten unless you keep the spirit and the enthusiasm alive. Organize your teachers and your children. Keep at it.

How may I secure permission from my principal to teach nature-study?

This inquiry I cannot answer, for it is a question of the personal point of view of the supervising officer, and possibly also of your own qualification. It is undoubtedly true that many good nature-study teachers are repressed and spoiled by principals, supervisors and trustees; but it is also true that many persons who think they can teach nature-study are self-deceived. Perhaps your superior has been prejudiced against the work by poor teaching on the part of some former teacher; it is scarcely possible that he could be now-a-days opposed to it on principle. If he is opposed on principle, there is probably nothing to do except to wait or to change your place. If he has had experience of shoddy work, you should ask him the privilege of giving a few lessons on trial, or should call his attention to the work or writing of a successful teacher. Perhaps your work with children at their homes would interest him. I think that most of the opposition to this teaching on the part of principals and superintendents is the result of misapprehension of what good nature-work is; it should be the pride of nature-study teachers to correct this feeling by doing the very best kind of work.

Would you teach heat, light and physics as nature-study topics?

Not as these subjects are ordinarily taught. They are usually taught as abstractions, having little relation to the pupil's life. There are many phenomena in these fields that are within the range of the pupil's experience, and these may be useful in the hands of a good teacher. The best results will be secured, by most teachers, by confining nature-study rather closely to biological fields and to those earth- and sky-subjects that are most intimately associated, in the child's mind, with the outside world. Many of the phenomena in this outside world are physical, and I would not exclude them; but I once knew a teacher who began nature-study for children with a disquisition on the conservation of energy!

Would you teach "practical" and "useful" things? (See pp. 90, 121, 129.)

Yes, if the things are such as appeal to the child and are adapted to the conditions. No, if they do not meet these requirements. In other words, I should not choose them merely because they are "useless" or "useful to man." I should want the child to have a wider horizon and a truer view of nature. The prime requisite is that the child become interested in the being itself, whether that being chance to be "injurious" or "beneficial." We must be careful not to dwarf the sympathies by purposely confining our work to those things that have "use." It is an error to assume that all the things in the world are important only as they relate to the financial profit and the pleasure of man.

On the other hand, I should not neglect the "practical" things just because they are practical and familiar. A horse, cow, pig, chicken, potatoes, wheat, cotton, alfalfa, and the rest, are excellent nature-study material, not only because they are intrinsically as interesting as other plants and animals, but also because they are common and therefore near to our lives. Familiarity should not breed contempt.

What one shall teach is determined very largely, of course, by the text-books in use in the school. The commonest fault that my informers find with text-books is that they have little relation to life; or as the persons themselves are likely to put it, the books are not "practical." I do not like to use this word "practical," because it has been employed in such a way as to arouse the antagonism of good teachers. Used in its original and legitimate sense it is well enough; but in order that the larger idea may be expressed, I like to say that text-books ought to be "applicable." The word practical is likely to connote merely dollars-and-cents information for the time being or for the place. The word applicable is more central, making the whole course of treatment, rather than a few isolated facts, significant to the life and interests of the pupil. The rigid text-book has been imposed on the schools by the colleges. With the emancipation of the schools, there should come a greater dominance on their part in educational policies. If the schools do not exist for the colleges, then it is very evident that a type of text-book that does not lead college-ward may be needed for the common schools; and this book will apply to the daily life.

Would you teach objects that the child cannot see and determine for itself?

No! Right here is where much of our nature-study effort shoots wide of the mark. The child should be set at those things that are within its own sphere and within the range of its powers. Much so-called nature-study teaching is merely telling the child what some man has found out. Bacteria, sheep's brains, complicated life-histories, chemical changes in germination, pollination, yeast, fermentation—these and a hundred others are beyond the child's realm.

How much apparatus do I need?

Perhaps none; possibly some. The apparatus and the method may easily be made too perfect. Any elaborate scheme or equipment is likely to be depressing to those who are less fortunately situated, if they are to teach. A laboratory in a teacher's training-school may be so extensive and complete that the graduates do not take up efficient work for themselves, feeling that they cannot do so without much equipment. Make the most of common and simple subjects, and leave the extensive outfits to teachers of science. Two pieces of apparatus that you ought to have are an aquarium for things that live in water and a terrarium for those that live on land. These become "scenes of life" and supplement the outdoors. (See p. 188).

Is it "thorough"?

"I do not believe in your nature-study movement," a high-school teacher said, "for it does not lead to thoroughness in school work." I asked her to explain what she meant by thoroughness. She took me to her schoolroom. It was a laboratory. Pupils of sixteen and seventeen were studying the cell. For three weeks the pupils had been working on the cell, and they were to continue the work for a month. This, she told me, was thoroughness. I agreed with her. "But of what educational value is this knowledge to the pupil?" I asked. "The pupil knows the cell," she replied, "and to know the cell is to understand the structure and growth of the plant."

We all believe in thoroughness, but there is one thoroughness of mere details and another thoroughness of the broader view. So far as mere thoroughness is concerned, one kind may be as perfect as the other. Thoroughness consists only in seeing something accurately and understanding what it means. We can never know all that there is to be learned about any object. Even the months' work on the cell was a mere smattering. Men spend their lives in studying the cell, and then do not understand it. What most school teachers mean by thoroughness is only drill in details. In its proper time and place, I approve this kind of drill in mere detail, but its place is not to dominate the school work.

But the great objection to my teacher's work on the cell, as I see it, is the fact that it means little or nothing to the pupil's life and is a mere acquirement.[115] We should put the child in contact with its own life, and the teacher who does this may teach with thoroughness whether he teach much or little. We can always be thorough and decisive as far as we go.

But will not this nature-study be called superficial? (See pp. 103, 132.)[116]

No doubt. A botanist told me that I was doing superficial work. Judged from the view-point of research, perhaps he was right; but I was not teaching science. Judged from the view-point of the child, I hope he was wrong. One is not superficial merely because he does not strike deep into subject-matter. He should try to be accurate as far as he goes. What is superficiality in the specialist may be commendable thoroughness in the layman. Even the specialist is satisfied with the most superficial knowledge in subjects outside his specialty. His knowledge of men and of business, for example, is likely to be superficial.

This charge of superficiality is usually only the opinion of a different point of view. This is well illustrated in the critical reviews of elementary text-books of science. Books that have been criticized severely by the scientist have been accepted with enthusiasm by the schoolmaster. The primary merit of a school-book lies in its pedagogy rather than in its science. Statements in such books have two values—the teaching value and the science value. Too often the reviewer thinks only of the science value.

Of course there is danger of superficiality. There is this danger in everything; but the danger is inherent in the person, not in the subject. Solid work is as necessary in nature-study as in anything else. It is not play, it is not sentimentality, and it is not blind wonder.[117]

Will not this nature-study tend still further to over-burden the school?

The overburdening of the school hours is due as much to the fact that the old subjects do not give way as that new ones are introduced. The old schools had too little variety. Perhaps the new ones have too much congestion. Just now we are in an intermediate stage between the old and the new. Nature-study is not a new subject demanding a place: it is a point of view asserting itself. It is an attitude toward life, and expresses itself in a way of teaching. Its spirit will eventually pervade and vitalize all school work.

It is some comfort to know that our school hours are now full. They cannot be fuller. If other things are added, old subjects must drop out. It is a struggle for existence. By introducing a freer treatment into some of the existing subjects, nature-study should relieve the congestion rather than increase it. If nature-study becomes a burden, it is likely to be because the teacher tries to teach too much and makes too hard work of it, or does not properly relate it to the other school work.

We still hear of many teachers who cannot find time to "introduce" nature-study; on the other hand we find many others, just as busy, who are able to flavor the whole school with it. If we accept that the nature-study spirit must be an attitude and a direction of thinking, then it does not at all follow that best results are to be secured merely by adding it as a separate period or task. The nature-study idea is something deeper and finer than simply another addition to the course of study, coördinate with customary school work.

We may need to take out subjects rather than put them in, and make every one of those that remain mean more. In time, the beginning schools will probably not teach any of the present-day subjects under their present names; but this will adjust itself in the natural course of evolution. The greatest need is to reorganize the teaching of the subjects that are already in the country schools.

Shall we teach the child to collect, and thereby to kill?[118] (See pp. 91, 106, 108, 111, 133.)

Properly directed, the collecting spirit should be encouraged, because one never comes closely into contact with his materials till he collects them with his own hands. To be close to one's material, develops enthusiasm and works itself into one's character. Every person should know the joy of finding something new.

How much or how little the collecting habit shall be encouraged must be determined for each case by itself; but, in general, the child should be taught to respect the life of every creature. Collecting should be an incident, particularly with very young children, and it should be encouraged only when it has some definite purpose. The spirit of savagery should be discouraged. I do not like to encourage young children to "catch things" for the mere excitement of catching them, but to study the habits of things as they are. I have little sympathy with the development of shallow sentimentalism regarding the life of animals and plants; but it is a safe principle, with children, to respect the life of everything, and to discourage the spirit of the hunter.

How may we develop the humane attitude toward living things?

In reply to your letter, asking how I would advise the teaching of "humane education" in the schools, I will say that I should let such teaching come as a result of a natural and well-directed development of the child. I should not teach tenderness, sympathy and morality directly as abstractions. I should try to interest the child in all living things, including other human beings, leading him to see their lives as they live them and enabling him to understand them. He then would have a reason for caring for them, and instruction would not be mere preaching (pp. 91, 145).

Of course, it does not follow that an understanding of the habits of animals and plants always insures humane feelings towards them, but if sympathy and spirit are a part of the teaching, it must inevitably lead in that direction. All first-hand contact with the verities of nature makes for ethical development of the individual.

Would you tell the child the names of the things?

Certainly, the same as I should tell him the name of a new boy or girl. But I should not stop with the name. Nature-study does not ask finally "What is the thing?" but "How does the thing live?" or "What does it do?" or "How did it get here?" or "What can I do with it?" The name is only a part of the language that enables us to talk about the object. Tell the name at the outset and have the matter done with (pp. 113, 133). Then go on to questions.

Would you begin by first reading to the child about nature?[119]

No, not in the school as a part of nature-study work. The reading should come after, not before (pp. 90, 93). Order will gradually come out of experience. The child should first come in contact with things rather than with ideas about things. This is the natural order. Animals come before zoölogy, plants before botany, fields and rocks before geology, words before language, religion before theology. Experience should come before theory.

There will be times, of course, in the exigencies of school work, when the teacher may feel obliged to read to the children in advance of taking up the particular study; but these occasions will be exceptions, and not a part of the system. In many cases, a vacant period or a rainy day may be made useful by good nature reading.

Now that there are so many nature-books, how shall I choose the most useful one?[120]

Only by finding out what you want. The multitude of books may be confusing, but the greater the number the greater is the chance that you will find one to your liking. Some persons deplore the making of many books, because they then have more difficulty in choosing; but the time has already passed when one book, or even two, can satisfy a good teacher. The teacher may not be able to purchase several books, but the school should supply a reasonable number. In these days the library is part of the equipment of the school. There is a general feeling that a new book—particularly a new school-book—is made for the purpose of displacing some other book. I once wrote a book. It seemed to occupy a field for which one of my best friends also had written. This friend wrote that perhaps I was right and he was wrong. I hope I was right but this does not imply that he was wrong. I hope that we are both right. There is more than one point of view.

It is not essential that we have uniform methods of teaching any subject in all parts of the country, and there is reason why we should not have them in nature-teaching. When one text-book satisfies everybody, it is because everybody is uncritical and unpersonal.

How shall I acquire sufficient knowledge to enable me to teach nature-study?

In the same way that you acquire other knowledge—by means of work and study. There is no way by which you can dream it or absorb it. There is no excellence without labor. The teacher should know more than he attempts to teach.

Yet, you must not magnify the importance of mere information. The ambition to teach and the love of doing for a child are the fundamental requisites.[121] Fill yourself full of some subject, however small it may be. When you cannot hold it longer, teach. Yes, you may make mistakes. But every one makes mistakes, even with the best of pains. Every person who, by teaching or writing, has helped the world to a higher plane, has said or written errors.[122] Every person, and particularly every teacher, should make all effort to be accurate; but if we wait till every possibility of error is removed, the world's work will never be done. Many a man sacrifices his chances of usefulness for fear of making a mistake.[123] The real work is not performed by timid persons (p. 103).

The best way to acquire the knowledge is to work for a time with a good teacher, who has enthusiasm and human sympathy. Read books and leaflets. Above all, go into the field and study the objects themselves. Do not wait until you are thoroughly equipped before you begin to teach, else you will never begin. When you have begun and your pupils begin to press for answers, you will learn. When you discover that you have made an error, admit it and acknowledge it. The pupil will respect you. Honesty always wins respect (pp. 96, 162).

It is not necessary that you become a scientist in order to teach nature-study. You simply go as far as you know, and then say to the pupil that you cannot answer the questions which you cannot. This at once elevates you in the pupil's estimation, for the pupil is convinced of your truthfulness, and is made to feel—but how seldom is the sensation!—that knowledge is not the peculiar property of one person, but is the right of any one who seeks it. It ought to set the pupil inquiring for himself. The teacher never needs to apologize for nature. He is teaching only because he is an older and more experienced pupil than his pupil is. This is the spirit of the teacher in the colleges and universities to-day. The best teacher is the one whose pupils the furthest outrun him; his pride is in the good pupils that he sends out.

Is it best to have a professional nature-study teacher to go from school to school?

This is a local, personal, and administrative problem. Ideally, it is best that every teacher handle the nature-study, because, as nature-study is a way of approach and a means of teaching, its effect is greatest when it is most continuous. In practice, however, some teachers will be sure to develop special aptitudes for the work, and these persons should be retained for this particular effort. The best talent should be employed for nature-study, as for anything else.

If there is a domestic science teacher going from school to school, perhaps she could also qualify in nature-study. Much of what we call domestic science is, or should be, pure nature-study; and all home questions should find expression in the schools.

Should not nature-study be in all the grades for all pupils, and technical work be left to the high-school?

[This teacher asks the following questions:

"Should not every teacher who goes out to the grades be prepared for giving the children instruction concerning the life about them? Should not nature-study be planned for all the grades as a means of giving the child his bearings and relations to animals and plants, and should not formal instruction in the principles of agriculture come in the high-school? or, in other words, should not the child's interest in things out-of-doors be fostered by means of informal and yet careful instruction during the earlier school years without special reference to the utilitarian phases of nature?"]

Your questions are easy for me to answer because they are framed in such a way that I need only to say "yes" to every one of them.

Nature-study teaching is not specialized teaching. It is a fundamental educational process which should put the child right toward the world and toward life. If every child should have a close connection with his environment, so, also, should every grown-up; and it follows that if the grown-up is a teacher, he will carry this spirit into the schoolroom.

The child who has the proper point of view toward the world in which he lives, and proper sympathy toward the objects and affairs about him, will be better prepared for any kind of study that comes later, whether that study is Latin, mathematics, engineering, agriculture, or other subject. I should leave the technical agriculture for the high-school, and preferably for the upper grades of the high-school. It is better to have the formal agriculture come after the student has had chemistry, physics and biology, at least to some extent. This would probably put the formal agriculture in the third or fourth year of the high-school. In the meantime, however, the pupil should have been prepared for all this work by having his mind open to the nature about him. In rural communities this nature-teaching will, of course, bring the child into touch with farms, whereas in cities and towns the farming phase of it would naturally be less emphasized. I should not try to force any child to become a farmer, or to follow any other occupation. When he comes to the realm of the high-school, he may of his own desire wish to begin to specialize. I should hope that the early training would be such that more persons would want to specialize in agricultural subjects than has been the case in the past; but the real nature-study teaching is quite independent of this.

It is undoubtedly a mistake to introduce formal and technical agricultural work into the grades. It is easy to refer the pupil in the grammar grades to bulletins and books, when he should be coming into original contact with the life and materials about him. The pupil should be taught to know domestic animals before he is instructed in the breeds of animals. He should know the way in which the neighbors build their houses and barns before he studies the styles of architecture. The grade work should touch many things, first and last, so that the pupil gains some conception of his world at large and, as you say, gets "his bearings and relations."

Should the parts of a school-garden be apportioned to pupils, or should the work be done in common?

In practice this becomes largely a question of administration: sometimes one thing may be done and sometimes the other. Ideally, the parts should be apportioned to pupils in the real laboratory school-garden. Thereby is the sense of proprietorship cultivated and the stimulus of emulation aroused. It is always advisable, when it can be arranged, to provide for some culmination or focus of the season's work in the nature of a flower-show or vegetable-show; or, the children may be allowed to sell the products of their gardens or to give them to hospitals or other worthy objects. This individuality of interest can be easily maintained in the plot-garden, but it is more difficult in the ornamental garden in which the plants are grown in continuous borders. (See p. 116.)

In order to indicate how some of the questions are attacked by those who are engaged in the work, I reprint an article on the Whittier School-Garden, by Miss Jean E. Davis, that appeared in *Country Life in America*:

"What is believed to be the largest school-garden in the United States is to be found in Virginia at the Hampton Institute for Negro and Indian youth, where it forms part of the equipment of the Whittier Training School—the practice-school of the institution.[124] Two acres of ground are given up to the garden, the larger part being divided into two hundred individual plots,

varying in size from four by six feet for the pickaninnies* of the kindergarten, to eleven by fifteen feet for the oldest boys and girls. Each plot is
owned, for the time being, by two children, who enter into partnership and
share equally in the work as well as in the profits of the garden—spading,
raking, planting, hoeing, harvesting with their own hands, and using the
products in their own homes or selling them to their neighbors. The young
farmers are not given *carte blanche*, however, in regard to the kind of crops
they shall raise or the position of them in the beds. The supervision of the
work is in the hands of one person—the director of the agricultural department of the Institute—who decides what vegetables and flowers shall be
planted and how they shall be arranged. This plan serves to give symmetry
and order to the garden as a whole, and adds materially to the educative
value of the work. Most of the plants selected are such as are easily cultivated and such as mature rapidly, like lettuce, radishes, nasturtiums and
marigolds; though peas, beans, cabbage, spinach and tomatoes are also cultivated. The gardens are made and planted both in the fall and in the spring,
the crops sown in the spring being cared for during the long summer vacation by volunteers.

"The beds are separated from each other by paths one foot wide, and are
arranged for the different classes in sections, having two-foot paths between
them. Extra plots, six feet wide, extending the full length of each section,
are used for overflow work by pupils who are exceptionally quick and energetic. Strawberries and raspberries are sometimes permitted in these beds.
Another opportunity for work out of the usual routine is afforded by a space
of three quarters of an acre which is reserved at the rear of the garden for the
purpose of teaching the larger boys how to use a horse and plow. In order
that the esthetic side of gardening may not be neglected—the cultivation of
a sense of beauty being esteemed of equal importance with practical instruction in agriculture—a large lawn has been placed at the entrance, while bor-
der beds of ornamental flowers form the other boundaries.

"But if school-gardening were confined to the making of gardens, the
planting of seeds and the cultivation of crops, beneficial as these experiences

---

* [Editor's note: In modern American usage, this is a derogatory slur referring to
young children of African descent. In 1903, the term was often used with inof-
fensive intent, like the phrase "colored children" that Davis uses later in the
article. However, as an othering and racializing term, it would always have been
perceived as derogatory by many. Consult the endnote to the previous sentence
for more on the legacy of nature-study programs like Whittier's.]

might be, it would still fall far short of accomplishing the end desired in introducing this subject into school courses. It would soon degenerate into either play or drudgery. To give it dignity and interest, and to make it of practical value in later life, the gardening is supplemented or preceded by simple experiments in the classroom illustrating the principles of germination and plant-growth; and a study is made of seed dispersion, the comparative value of soils and the work of beneficial and injurious insects. Seeds are planted in window-boxes, the seedlings affording material for language and drawing lessons before being transplanted into the outdoor beds. The decorative value of flowers, leaves and berries is considered, and the children are encouraged to make gardens at their homes from which they may gather bouquets of flowers for their dinner-tables.

"The results of two years' experience in teaching gardening and nature-study at the Whittier School are most gratifying. While at first it was necessary to use compulsion with some of the older girls, and the little ones merely considered anything 'good fun' that took them out of doors, they now without exception look forward with eager enthusiasm to 'gardening day,' which comes twice a week to each of the four hundred. Large crops have been gathered and proudly carried home; seeds have been in demand for home gardens, sixty or more of which have been made in the neighborhood; and last spring children to the number of one hundred and thirty volunteered to cultivate the gardens during the summer vacation. In the home-gardens there has been great diversity of crops. Besides the usual school plants, children have raised wheat, corn, pumpkins, sweet and Irish potatoes, and also many kinds of flowers. A wholesome rivalry has sprung up between the owners of adjoining beds in the school-garden, and pride in the appearance of the school-grounds has been stimulated. An interest in birds and insects, and an appreciation of the beauty of wayside flowers and other common things, have been developed; and the roughest children have been made more gentle by handling the beautiful flowers that they have grown, the result of their own care and patience. A regard for the property and rights of others is among the results of this coöperative gardening, also an appreciation of the advantages of working together, and a certain forbearance and loyalty to one's partner, all of which are lessons of inestimable value, especially to colored children. When we add to these unconscious influences of school-gardening the conscious self-respect and self-reliance that come from the ability to produce from the soil something of one's very own, it will be admitted that this subject is worthy of an honorable place in the course of study of our common schools, of which the Whittier School is only a type."

Can I make a nature-study exhibition useful as a part of an exposition?

I hope to see good nature-study exhibitions at all the great expositions. It is time that we begin to relate education directly to the affairs of life; or, to put the matter in another way, to make the affairs of life a means of education. I hope that you will find some way of making your educational exhibition dynamic. Most exhibitions are merely passive or static, consisting of pictures and charts, books, apparatus, and such other things as sit still. The very essence and spirit of the new education is activity. I judge from your letter that you are expecting to express this activity by means of a school in actual operation. I hope that you may also have a good school-garden in actual operation, and also some effective outdoor laboratory work. I am not yet satisfied with the school-garden movement. I think that we have not yet developed its laboratory significance.

The time is coming when we shall begin our educational process by putting the child into real activities of work and play, and when we shall add the books and apparatus gradually as he grows and the need of them develops. Your exhibition should teach this.

Should this nature-study be confined to the schools?

It should not be confined to schools. Too often it is thus restricted because we are in the habit of delegating the training of our children to a professional class of teachers. Ideally, the home should be the most perfect school, and the parents should be the best teachers. In the increasing complications of our lives, however, the division of labor forces the children more and more from the home-training into the school-training; therefore it is increasingly important that we give good heed to the maintenance of schools. But even so, the home-training should afford an auxiliary to the school-training. There should be more than one common bond of method and purpose. One of these bonds should certainly be the desire to put the child into sympathetic relation with its own necessities.[125]

I fully commend education by means of literature and history and science and art, of course; but if I were confined to one means I should choose that which would lead me to love the things that I see and the work that I do day by day. This outlook I should want to impress on my children;[126] but I could not impress it by any mere intellectual means. It is an affair of the heart; and if I do not live it I cannot teach it.

But it does not follow because one or even both of the parents is in full rhythm with the natural world, that the parents can teach the child effectively. Few persons are good teachers; and when there is marked difference of outlook between the parents, the school may be the only agency that can give the child an harmonious relation.

The school is a distributing agency for all kinds of educational ideas. It must more consciously recognize this function and take pains to aid parents, pastors, and others to encourage good work outside the school, particularly such work as contributes to the prosperity of the community. The high-school bears a marked responsibility in this way, because it has greater equipment than the grade-school and deals in more particularized subjects. The influence of the high-school should be felt not only in the school grades, but in the whole daily life of the people. It should set good ideals of public service by enabling the people to meet their problems.

What shall we do with the children in the summer vacation?

This is an exceedingly important question and very difficult to answer. The teacher has no control of the child during this period. He can suggest what the pupil may do, but the probability is that the pupil will merely drift.

I am convinced that there is a great loss of efficiency in the over-long and undirected summer vacation for both child and youth. The colleges are beginning to feel this, as shown in the development of four-term systems. The summer schools are protests against an idle summer. Herein is where the farm boy acquires much of his efficiency for the battle of life—in the fact that he has no long periods of enforced idleness, laziness and emptiness. He is kept at work. He grows up with an appreciation of the value of time. He knows what industry is and what it brings. Steady effort and application become the warp and woof of his life. The town boy of the upper and middle class, on the other hand, is likely to become accomplished in feats of idleness. One fourth his time is mere vacation, or, rather, mere vacancy. He is handicapped when later he comes squarely against the realities of life.

I believe in a long vacation if the time is occupied in some well-directed effort. I am glad to see the development of the summer-camp idea for both boys and girls, where, under competent and sympathetic guidance, with firm but kindly discipline and something like Spartan fare, they are led to see and to know the nature in which they are.[127] In such camping-out experiences the youth comes hard against actualities. He gathers materials that are his own and that become a part of his capital throughout life. He comes to his own conclusions and to think for himself, not merely to absorb his knowledge and opinions from teachers and books. In later life he may never have another opportunity to secure this actual experience.

I wonder how many persons ever saw the sun rise?

Will not this nature-study work interfere with school discipline?

That all depends on what you mean by "discipline." If you mean perfect "order," the child sitting erect with clasped hands, then nature-study work may annoy you. If you mean only that the child is well-behaved, obedient and happy, then no ill result should come from the nature-study effort. Nature-study should supply some of the "busy work" between the regular periods. The best means to secure good discipline is to keep the child busy and interested. "Discipline" is then a result.

The greater number of mischievous and refractory children can be interested in some piece of personal work or investigation. The boy who is "licked" at home and punished at school is likely to spend his time midway between the two; and yet he may be easy to reach if only he is understood.

Shall I correlate the nature-study work with other work?

This question can be answered only for particular cases. In general, correlation is an advantage to all subjects concerned; however, I fear that in much of the correlation the nature-study part is little more than a name. If the nature-study can be kept genuine—a real study of native objects and relations at first hand—I see no danger in correlation. The correlation usually is of greater benefit to the other subjects than to nature-study.

Nature-study work can be correlated with various other school work, notably with essay writing, drawing and geography teaching. The very first essential in essay writing is to have something from one's own experience to say. Assigned topics are usually "hard" at best. Let the child write of what it has seen or done that day or yesterday—the butterfly, the tadpoles in the pond near by, the plants growing in its garden, the fish in the aquarium, the peaches on the tree by the barn, the little world of life in the terrarium, the woodchuck that lives under the stone fence, the things it saw in the market, the vehicles it sees on the street, the factories and farms near by, the field work, the house work, the school, the highway, the hill, the kinds of fences by the way, the collecting expeditions and the games. If the child has had no such experience, why not begin by assigning him a living topic to look up and report on in writing?

We need to be unusually careful to see that the writing is not exotic to the child. Avoid the model of nature-study "stories" and "write-ups" about things; these stories tell what others have found out. They may inform and instruct and entertain, rather than educate and set the child to work.

We stifle the desire to write if we first lay down rules and formulas as to how to write. Let the child have a personal experience; then allow it to write. Did you ever have a pupil who could not write a composition, but who could write a letter that was full of originality and personality? Why could it write the one and not the other? Too often, I fear, we prevent our children from writing by trying to make them write. Of what use is writing, anyway, if it is not self-expressive? So, let the child have something real and personal to write about. No subject is too mean. Then when the child has written, throw away the blue pencil and suggest tactfully how the piece may be improved here and there. Do not hinder the child.

I well remember my first "composition." For days I had tried to think of a "subject." I had importuned father and mother and friends. "Winter," "Spring," "The pen is mightier than the sword," "The pleasures of farm life," "Shakespeare"—all had equal terrors. Rapidly the days passed away, and to-morrow the composition must be ready, and yet of all the well-sounding subjects not one seemed to present a way of escape. The teacher—God bless her!—learned of my plight. She asked me what was the best "time" I had had last summer. Of course I knew—the time when we all went blackberrying, with all of us rolled into the bottom of the wagon-box that went bumping and rattling over the stones and grinding through the sand, when we crept through the deep cool woods and then came into the "clearing" where the skidded logs were covered with the tangle of berries and berries—of course I knew! With what wild delight I told her! and then she said, "Just write that down and it will be your composition." From that day until this I hope I have written only on those things that are dear to me.

I have a similar word to say about drawing. The other day I heard Mrs. Comstock[128] speak on this subject before a convention of teachers. She is herself an artist. She said that there are two kinds of drawing—the kind that is the child's self-expression, and the kind that makes an artistic picture. It is natural for every child to make lines and marks to express what it sees or experiences; but when these lines and marks do not conform to the ideals of grown-ups, we discourage the effort and the child ceases to draw. Considered as the effort of the child to express itself, no drawing can be "poor." Mrs. Comstock put on the board a copy of a drawing from a child's pad, and it was as follows:

Figure 9. How a man impressed a child.—face, arms, legs

We all laughed; but we were told that this was no caricature, but the impression that a man made on the child—face, arms, legs.

More than words, the drawing may show what the world means to the child, even allowing for all the errors in clumsiness with pencil. Do you not wonder how the world looks to the little girl in the second grade who made all these drawings and sent them to Uncle John? Would you not like to take her on your knee and have her explain them to you?

Figure 10. What a little girl saw

Primarily, drawing is a means of expressing what we see and feel; now and then a person develops the ability to make a picture that pleases others, and he becomes an artist. Primarily, our interest in the external world is one of sympathy and personality; now and then a person develops the ability to make discoveries and to record them, and he becomes a scientist.

Correlation of nature-study and drawing should give excellent results to both subjects. The nature-study should afford objects in which the pupil is genuinely interested; the drawing should aid in focusing the observation and making it accurate. Drawing should be encouraged primarily for the purpose of discovering what the child really sees. As the child sees more, and with greater accuracy, the drawings improve. So the drawings become an approximate measure of the progress of the pupil. Do not measure the drawings merely as drawings, or from the artist's point of view. We are likely to dwell so much on the mere product of the child's work that we forget the child.

Too early in the school life do we begin to make pupils mere artists and literators. First the child should be encouraged to express himself; then he may be taught to draw and to compose.

If correlation produces these useful results, it should be encouraged.

What can I do to put our rural schools in touch with their constituency?

What you can do, as a superintendent, to aid your rural schools to better their conditions, is to enter into a general agitation of the subject through the local papers, through correspondence with the teachers themselves and the school officers, with the granges,[129] and other farmers' societies, village improvement societies, pastors, and whoever and whatever else there may be that stands for bettering conditions.

Work of this kind cannot be accomplished in any one way or through any one source. With a determination to alleviate the situation, with imagination and with industry a person can accomplish a good deal in the directions about which you inquire.

Following are definite suggestions to make to rural teachers for working out in the schoolhouse (adapted from M. P. Jones, Cornell Rural School Leaflet):[130]

1. *Register* with the college of agriculture or experiment station of your state, to receive the publications and to be on the correspondence lists.
2. Write to the state education department for whatever *syllabi* it may publish on nature-study, agriculture, and similar subjects.
3. Start an *agricultural and nature-study library*. A very creditable beginning may be made at no cost except postage, by asking for publications issued by the Department of Agriculture at Washington and the State Agricultural College and Experiment Station. It is recommended

   (*a*) That you write to the Department of Agriculture, Washington, D. C., asking to have the school placed on the mailing list for the monthly list of publications and to have the following sent to you:

   > 1 set of Farmers' Bulletins suitable to the locality.
   > 1 copy of the list of Publications for Free Distribution.
   > 1 copy of the list of Publications for Sale.
   > 1 copy each of reprints of areas that have been surveyed by the Bureau of Soils in your state.
   > 1 copy each of Bulletins 186 and 160 and Circulars 77 and 52 of the Office of Experiment Stations.[131] On receipt of these bulletins, holes should be punched through them and strings used to tie them together. Manila paper may be used for covers.

   (*b*) That you write to the Geological Survey, Washington, D. C., inclosing 15 cents in stamps and asking for the three geological survey maps that cover your region.

   (*c*) That you write to your representative, congressman or senator, for copies of state or national documents that are distributed by them.

   (*d*) That you secure the use of a traveling library, if such libraries are issued by the education department or other agency in your state.

   (*e*) That as many agricultural and nature-study books be added to the library as money will permit.

4. Beautify the *school-grounds*. Endeavor to interest the trustees and the farmers of the district. In one district, an oyster supper brought forth money and enthusiasm enough to produce a marked improvement. Valuable suggestions will be found in

Farmer's Bulletin (U. S. Dept. Agriculture) No. 43, "Tree Planting on Rural School Grounds."
Farmer's Bulletin No. 185, "Beautifying the Home."
Farmer's Bulletin No. 248, "The Lawn."
Farmer's Bulletin No. 218, "The School-Garden."[132]

Perhaps the agricultural college of your state, or your state education department, has issued publications on this subject.

5. Begin a *school-garden*. Every country school should have its garden. If possible, it should be large enough so that every child may have a garden of its own. The children should also be encouraged to have gardens at home. The school-garden may be used as an experiment station to test fertilizers, varieties, methods of planting, and the like. Read books and bulletins on the subject.

It will be found less expensive to buy seeds in bulk and divide these into penny packets to be sold or given to the children, preferably sold.

6. Make a *window-box* and have plants growing in it.
7. Have a *terrarium*. This is a box with sides and top made of window screens. The top is hinged so that it can be raised. Earth may be put in the bottom and plants allowed to grow in it. Frogs, toads, insects and other outdoor life can thus be safely housed. The terrarium may be used in winter in the study of fowls.
8. Have an *aquarium*. A glass vessel or a Mason fruit-jar, with water frequently renewed, will serve for a time. Have some water-plants growing in the aquarium and keep a few fishes, salamanders and tadpoles for study.
9. Have a *museum* of things related to the life and affairs of the region.[133] Let the collection be started and increased by the children themselves. It is suggested that collections be made of the following:

   (*a*) The different types of soil found in the neighborhood: sand, silt, clay, muck, and sandy, silty and clay loams.

(*b*) Seeds of common vegetables, flowers, fruits, and trees.

(*c*) Common grasses: timothy, red-top, meadow fescue, Kentucky blue-grass orchard-grass.

(*d*) Common legumes of the farm and garden: red, white, and alsike clovers, alfalfa, peas, beans, vetch, soy beans, cowpeas.

(*e*) Common cereals: corn, wheat, oats, rye, barley, buckwheat, rice.

(*f*) Ears of corn: flint, dent, pop, sweet. Secure ears showing the qualities that good ears should have. A lesson in corn-judging may profitably be given.

(*g*) Fertilizers: nitrate of soda, dried blood, ground bone, acid phosphate, muriate of potash, and as many others as are used in the neighborhood.

(*h*) Feeds for farm animals: bran, middlings, gluten feed, buckwheat middlings, and others in use. The local feed merchant and seedman might lend their aid in supplying samples of these feeds, samples of fertilizers and seeds.

(*i*) Fruit. In the fall, different varieties of apples, pears, plums, and grapes could be collected, probably with much enthusiasm, by the children. Part of an afternoon could be given for a short talk on fruit-growing by a local fruit-grower, after which the samples of fruit could be eaten. Similar collections of root-crops and vegetables might be made, not with the idea of keeping them in the school for a long time, but as one of the best means of teaching children to become familiar with the common things of their farms.

(*j*) Flowers and weeds. These can be pressed and used as the basis for the school collection. Begin with the most common plants and enlarge the collection slowly.

(*k*) Leaves of trees. Press the leaves of some of the most common trees, adding to the collection slowly enough for the children to learn as they go.

(*l*) Fibers: wools of different kinds, cotton, flax, hemp; ropes, twine (particularly binder twine), bagging, fabrics, etc.

10. Teach the *Babcock milk test*. Some schools have demonstrated the use of this test before grange meetings. Complete milk-testing outfits suited for school use are manufactured at small price. Write to dairy supply house for catalogues, and get information from your college of agriculture.

11. Have a *reading-table*. Secure a few good magazines, agricultural and other kinds. No poor books or poor magazines should be in the schoolroom or home. Some publishers of agricultural magazines will send complimentary copies if, in asking for them, it is stated that they are wanted for the school library.

12. Have a *work-bench* with tools, if possible. The boys and girls should become familiar with the handling of common carpenters' tools. Simple things, especially those that can be used on the farm or at play, may be made, such as a window-box, terrarium, stakes for the school-garden, bird-houses, kites, sleds, skees, book-shelves, tables, flower-stands. Hand tools can be repaired. This will provide excellent manual-training, developing naturally into use of wheels and more complex forms.

13. Have one or two *vases* with flowers well arranged.

14. Have a *school fair*. These have been found very successful where tried. Children exhibit products from their own gardens and benches. The girls exhibit cakes, pies, biscuits, which they have made. Small prizes are given. The people of the district are invited and the fair is made one of the important social events of the year. It will probably be found that the older people enter enthusiastically into a competition of their own, and if this can be arranged it will add greatly to the success of the fair. Take the exhibit to the county fair or state fair.

15. Take *occasional trips* to neighboring farms, factories, to the woods and fields.

16. Provide some simple *apparatus*, as, for example, the following to begin with:

    1 Babcock milk test (if in a dairy country)............................ $5.00
    1 tripod lens magnifying glass .... .75
    1 terrarium.. 1.25
    1 aquarium .... 2.00[134]
    1 insect net (home made)
    Various cups and boxes to hold specimens.

17. Try to know the *weather*. If you have expeditious mail service, apply to the United States Weather Bureau for the daily bulletins and a frame to put them in. A good thermometer should be hung in a

protected shady place. Thermometers that are reliable at high and low temperatures usually cost more than one dollar. A rain-gauge will be useful and interesting. Some schools may add a barometer, if the teacher understands it; but the cheap instruments are not reliable.

How can I reach the farmers of my neighborhood?

[A teacher is discouraged because she seems to make no headway; and the farmers complain that her work is not practical and they want to know how to make more money.]

While you are under obligation to teach farmers' children, you do not bear the responsibility of making the farms profitable. It is the business of the farmer himself to make his farming pay. You are engaged in the work of education.

How to teach, not how to farm, is therefore your problem. I take it to be axiomatic that every person's mind should be expanded in order that he may derive the greatest satisfaction from life. If the occupation in which he is engaged will not allow him to derive this satisfaction, then it is his privilege, and in fact his duty, to change his occupation. I am very sure that the educating of farmer boys and girls will often have the effect of taking them away from the old farm. It is a question, then, whether the whole point of view on farming must not change and whether such new methods and new types of life must be developed as to interest persons with a broad outlook on life. I think that the diffusion of information and the extension of education is bound to have this effect on the farming industry in the long run. In the meantime, it is for us to try to determine just what is the most practicable means of procedure in the educating of the country boy and girl, that will give them a satisfactory outlook on life, and make them least willing to give up their place in the country.

Time and again I have had problems similar to the one that your patron asks of you, namely, that instead of giving scientific information about eggs, you tell him how to make his hens lay better when eggs are scarce. It is very easy to ask how to make hens lay in October and November; it is quite another thing to answer the question. Such a question cannot be answered out of hand. A man must first learn something about breeding, and feeding, and care, and other things. In other words, a man must have enough fundamental knowledge to know the reasons why, and this knowledge is necessarily scientific. It is utterly impossible to try to answer the greater part of our agricultural questions until the questioner has some really underlying understanding of the conditions, and processes, and principles involved. The lack of this understanding is one reason why farmers are so backward in utilizing advice, and also why they are unable to use the experiment station bulletins.

But even if you could tell your patron how to make hens lay in October, that would not settle or simplify your teaching. You must lead your pupils to go beyond an isolated fact and relate it to other facts. You must give them some conception of the hen's habit of life. You must not allow your advice to farmers to take the place of the training of farmers' children.

I do not doubt but that all elementary educational work for country conditions is yet very crude and fails adequately to reach the mark. On the other hand, I am convinced that we are learning how. In the meantime it seems to me that it is your part as a teacher to endeavor to put the country children, as much as possible, directly into touch with their environment in order that they may understand it and appreciate it. I am quite sure that not all the compensations of farming are in the shining dollars of which you speak. Some of the compensation comes in a sympathetic appreciation of the surroundings and the advantages that a farmer has and may have; and the countryman cannot be really successful until he arrives at this appreciation. Of course, he must first of all have the money, for this enables him to live; but there are other rewards in life. If the farmers do not appreciate all this, you must do your work just the same, and wait.

You certainly are not alone in feeling that you cannot carry the children much beyond the printed lesson. As you say, these subjects are so new that there has been no opportunity for adequate training in them. I think that the best teacher I ever had along these lines was a woman who knew very little about the subject-matter itself, but who encouraged me, answered my questions as best she could, and told me frankly when I had found out more than she knew.[135] I judge, however, that you quite underestimate your own knowledge, else you would not feel so keenly the responsibility of your work.

You can do a great deal outside the school for your people. You can work through farmers' organizations, attend farmers' institutes, help to organize boys' and girls' clubs, reading clubs, help to put educational work in the fairs, and in many other ways quicken the rural life of your vicinity.

How can a teacher prepare himself to teach agriculture in the special schools that are now being established?[136]

Beyond pleasing personality and moral character, there are two powers that qualify a person to teach: (1) the teaching ability, which is in part a natural quality and in part gained by experience; (2) knowledge of the subject-matter.

The subject-matter can be acquired partly by attendance at summer schools and by home reading, but if you are intending to fit yourself for the best positions you will need to attend a good college of agriculture. Even though you are farm bred and know the practical business of farming, you will need the college training to give you a rational grasp of the field and to enable you to put your abilities into teaching form. For these best positions, you must take nothing less than a full four-year course, for you will have to compete with the regular graduates of these institutions; and four years' training is little enough to fit you in the fundamental sciences and arts, and to prepare you in the modern agricultural subject-matter. For those who cannot take full training, the colleges of agriculture offer short and special courses.

[I have given a full outline statement of these questions in Bulletin No. 1, 1908, of the United States Bureau of Education, under the title, "On the training of persons to teach agriculture in the public schools."][137]

How can I do any nature-study work in the ordinary kind of schoolroom?

School buildings are constructed for the work that is known and recognized at the time of their erection; so it follows that they may be very poorly adapted to nature-work. If your room or building is poorly adapted, you will be obliged to shift as best you can, making the most of unsatisfactory conditions. You should not give up the work for that reason. You may have room at one side or end for a table on which you can place a terrarium and aquarium and other things. You may have a window or two in which it will be possible and advisable to grow plants. In some cases, the children can germinate a few plants, or even raise them to maturity, on their desks. You may have a yard in which a little can be done in gardening. If you have none of these possibilities, then you can encourage the pupils to grow plants and to make their observations at home (which they should do anyway) and report the results in school. You can have them bring in such specimens as do not require to be kept, and then "clean house" frequently.

In the planning of new school buildings, ample provision should be made for nature-work. The need of this is particularly apparent in the country schoolhouses. In rural districts, we must have a new kind of schoolhouse. A room or wing should be added for work with tools and with nature objects; or a basement may be provided; or, in many district schools in which the number of children has decreased, one end of the old schoolroom may be partitioned off for this purpose; or some good outbuilding may be requisitioned. The school premises of the new order must be provided with good grounds, and these grounds should grow many or most of the native trees and shrubs of the neighborhood, becoming a little local park and a beauty spot.

We have talked much about new teachers, but we need schoolhouses about as much as we need new teachers. I suppose they will come together. There is no use of evading the question of better equipment. We must put more money into our schools if we expect to make them better. Schools are worth about what they cost. We must not only have new pieces of equipment, but a wholly new idea of equipment. We are to go back to the beginning and do it all over again and begin naturally and practically. Different kinds of things must be put into schoolhouses from those that we have been accustomed to put there (pp. 187–191). We must put in them products and implements, and make them express the life and enterprises of the neighborhood. We must improve not only the school and premises but we need equally to interest the whole district or constituency in the better things.

It is not the teacher alone or the schoolhouse alone that we need to improve. We have talked about the little red schoolhouse; but the little red schoolhouse (as one of my farmer friends puts it) is likely to contain the little green teacher.[138]

Is nature-study on the wane?

Real nature-study cannot pass away.[139] But the more closely we come into touch with nature the less do we publish the fact abroad. We may hear less about it, but it will be because we are living nearer to it and have ceased to feel the necessity of advertising it.

Teaching may not be nature-study merely because it is so called. A superintendent told me that he had forbidden nature-study in his schools. I asked him what the work had been. He said that it was the dissecting of cats. A publisher told me that nature-study is passing out. I asked why he thought so. He replied that his nature-study books were not selling as well as they did. I told him that I was glad.

Much that is called nature-study is only diluted and sugar-coated science. This will pass. Some of it is mere sentimentalism. This also will pass. With the changes, the term nature-study may fall into disuse; but the name matters little so long as we hold to the essence.

All new things must be unduly emphasized, else they cannot gain a foothold in competition with matters that are established. For a day, some new movement is announced in the daily papers, and then, because we do not see the headlines, we think that the movement is dead; but usually when things are heralded they have only just appeared. So long as the sun shines and the fields are green we shall need to go to nature for our inspiration and our release; and our need is the greater with every increasing complexity of our lives.

Would you advise me to take up nature-study teaching?

Yes, if you feel the "call" to it; otherwise, no. I would have only those teachers teach nature-study who are well qualified for it, as I would advise for grammar or other school work. Every teacher ought to have the nature-study outlook to keep him young and interested in life, but we all recognize that relatively few of them have it. Every pupil should have nature-study, under one name or another; but he should receive his inspiration from the teacher who himself is so full of the subject that he teaches with spirit and with cheerfulness.

After a time, we shall not need to argue for nature-study. Teaching must in the end be natural.

Major Sections Restored
from the First Edition

FROM PART I, CHAPTER VII

The Agricultural Phase of Nature-Study

This chapter was largely rewritten for the third (1909) edition, under the new title "Nature-Study Agriculture," as it appears in this volume, with only a few paragraphs surviving from the first edition. While the revision brought the chapter more in line with the tone and content of the book as a whole, focusing more on the general philosophy undergirding nature-study's relationship with agriculture, the first edition of the chapter gives us a much more detailed glimpse into the fuller scope of Cornell's nature-study work in New York and the specifics of how it was being implemented. It also demonstrates how that work related to Bailey's larger vision for what would become the country life movement and what Bailey here calls "the spiritualizing of agriculture"—a counterforce to the efficiency and output-oriented changes he saw happening in agricultural education as the country reacted to the economic panics of the late nineteenth century. Scans of the numerous specific bulletins and leaflets that Bailey names in this section are available through the digital companion exhibition to this book at www.lhbaileyproject.com. The following passage is presented as it appeared in the first edition. It picks up at the end of the chapter's second paragraph (the first two paragraphs remaining roughly intact across editions), which in the first edition concluded with the following extra sentence, and it continues to the chapter's original conclusion:

Yet, as a matter of fact, what do our rural schools teach?

So long have we taught the text-book routine that we do not seem to think that there may be other and better means. I believe in the Greek idea of education for culture, but I would have other education along with it. I believe that it is possible to acquire culture at the same time that we acquire power. Education for culture alone tends to isolate the individual; education for sympathy with one's environment tends to make the individual an integral part of the activities and progress of its time. At all events, I cannot see why there is not as great possibility for culture in the nature-studies as there is in the customary subjects of the elementary school. My plea is that new educational methods must be employed before we can really reach the farming communities. Nature-study is to supply some of these new means. Nature-study must be made a part of the extension-teaching of the time—of that movement which takes the school to the people when the people will not go to the school. The educational impulse must be taken to every man's door. If he shuts the door, it must be thrown in at the window.

All agricultural educational work is yet in an experimental stage in this country, with the single exception of college work—and even this is likely to be much modified within the next few years. Therefore, there are no perfect or generally accepted methods of nature-study as applied to rural education; but sufficient experience has now accumulated to enable any good teacher to make a beginning anywhere with full assurance of doing useful and lasting work. The direct application of nature-study to agricultural education appears to have been started by the Agricultural College of Cornell University. This was in 1895 and 1896. This work is of a true extension character, being conducted from the university as a center, by means of lectures, publications, correspondence, and the organizing of pupils into clubs. It is advisory and propagandic. Its object is to interest teachers and pupils of the public schools in nature-study work with special reference to the agricultural conditions. The first necessity in the work proved to be the need of instruction for the teacher; and to meet this necessity special literature was prepared in the form of "nature-study leaflets." These are designed to inspire the teacher, to give him point of view, to send him directly to nature to verify the facts and to extend his knowledge, to suggest methods of teaching the subjects. They are not texts from which recitations are to be made. Merely as an example of one set of ideals and one method of improving the agricultural status, a brief outline of this work may be given. The following extract is from a sketch which I contributed to the Sixth Report of Extension Work (Bulletin 206, Cornell Experiment Station, October, 1902):

"To create a larger public sentiment in favor of agriculture, to increase the farmer's respect for his own business—these are the controlling purposes in the general movement that we are carrying forward under the title of nature-study. It is not by teaching agriculture directly that this movement can be started. The common schools in New York will not teach agriculture to any extent for the present, and the movement, if it is to arouse a public sentiment, must reach beyond the actual farmers themselves. The agricultural status is much more than an affair of mere farming. The first undertaking, as we conceive the problem, is to awaken an interest in the things with which the farmer lives and has to do, for a man is happy only when he is in sympathy with his environment. To teach observation of common things, therefore, has been the fundamental means. A name for the movement was necessary. We did not wish to invent a new name or phrase, as it would require too much effort in explanation. Therefore, we chose the current and significant phrase 'nature-study,' which, while it covers many methods and practices, stands everywhere for the opening of the mind directly to the common phenomena of nature.

"We have not tried to develop a system of nature-study nor to make a contribution to the pedagogics of the subject. We have merely endeavored, as best we could, to reach a certain specific result—the enlarging of the agricultural horizon. We have had no pedagogic theories, or, if we have, they have been modified or upset by the actual conditions that have presented themselves. Neither do we contend that our own methods and means have always been the best. We are learning. Yet we are sure that the general results justify all the effort. In fact, we never believed so fully in the efficiency of this kind of effort as at the present time.

"Theoretical pedagogic ideals can be applied by the good teacher who comes into personal relations with the children, and they are almost certain to work out well. They cannot always be applied, however, with persons who are to be reached by means of correspondence and in a great variety of conditions, and particularly when many of the subjects lie outside the customary work of the schools.

"Likewise, the subjects selected for our nature-study work must be governed by conditions and not wholly by ideals. We are sometimes asked why we do not take up more distinctly agricultural or economic topics. The answer is that we take subjects that teachers will use. We should like, for example, to give more attention to insect subjects, but it is difficult to induce teachers to work with them. If distinctly agricultural topics alone were used, the movement would have very little following and influence. Moreover, it is not our purpose to teach technical agriculture in the common schools, but to inculcate the habit of observing, to suggest work that has distinct application to the conditions in which the child lives, to inspire enthusiasm for country life, to aid in home-making, and to encourage a general movement toward the soil. These matters cannot be forced. In every effort by every member of the extension staff, the betterment of agricultural conditions has been the guiding impulse, however remote from that purpose it may have seemed to the casual observer.

"We have found by long experience that it is unwise to give too much condensed subject-matter. The individual teacher can give subject-matter in detail because personal knowledge and enthusiasm can be applied. But in general correspondence and propagandist work this cannot be done. With the Junior Naturalists,[1] for example, the first impulse is to inspire enthusiasm for some bit of work which we hope to take up. This enthusiasm is awakened largely by the organization of clubs and by the personal correspondence that is conducted between the Bureau and these clubs and their members. It is the desire, however, to follow up this general movement with instruction in definite subject-matter with the teacher. Therefore, about a year ago a course in Home Nature-study was formally established under the general direction of Mrs. Mary Rogers Miller.[2] It was designed to carry on the experiment for one year, in order to determine whether such a course would be productive of good results, and to discover the best means of prosecuting it. These experimental results have been gratifying. Nearly 2,000 New York teachers are now regularly enrolled in the Course, the larger part of whom are outside the metropolitan and distinctly urban conditions. Every effort is made to reach the rural teacher. Plans are now making for the modification of this Course, by means of which it is hoped that the number of teachers receiving definite correspondence instruction will be very largely increased. [The number has now reached nearly 3,000, February 28, 1903.]

"In order that the work may reach the children it must be greatly popularized and the children must be met on their own ground. The complete or ideal leaflet may have little influence. For example, I prepared a leaflet on 'A Children's Garden' which several people were kind enough to praise. However, very little direct result was secured from the use of this leaflet until 'Uncle John' began to popularize it and to make appeals to teachers and children by means of personal talks, letters and circulars. So far as possible the appeal to children was made in their own phrase. The movement for the children's garden has now taken definite shape, and the result is that more than 26,000 children in New York State were raising plants during the present year. Another illustration of this kind may be taken from the effort to improve the rural school-grounds. I wrote a bulletin on 'The Improvement of Rural School-Grounds,' but the tangible results were very few. Now, however, through the work of 'Uncle John' with the teachers and the children a distinct movement has begun for the cleaning and improving of the school-grounds of the State. This movement is yet in its infancy, but more than 400 school-yards are now in process of renovation, largely through the efforts of the children.

"The idea of organizing children into clubs for the study of plants and animals and other outdoor subjects, originated, so far as our work is concerned, with Mr. John W. Spencer, himself an actual and practical farmer.[3] His character as 'Uncle John' has done much to supply the personality that ordinarily is lacking in correspondence work, and an amount of interest and enthusiasm has been developed amongst the children which is surprising to those who have not watched its progress.

"The problems connected with the rural schools are probably the most difficult questions to solve in the whole field of education. We believe that the solution, however, cannot begin directly with the rural schools themselves. It must begin in educational centers and gradually spread to the country districts. We are making constant efforts to reach the rural schools themselves, and expect to exhaust every means within our power, but it is work that is attended with many inherent difficulties. We sometimes feel that the agricultural status can best be reached through the hamlet, village and some of the city schools rather than by means of the red schoolhouse on the corner. By appealing to the school commissioners in the rural districts, by work through teachers' institutes, through farmers' clubs, granges and other means we believe that we are reaching farther and farther into the very agricultural regions. It is difficult to get consideration for purely agricultural subjects in the rural schools themselves. Often the school does not have facilities for teaching such subjects, the teachers often are employed only for a few months, and there is frequently a sentiment against innovation. It has been said that one reason why agricultural subjects are taught less in the rural schools of America than in those of some parts of Europe is because of the few male teachers and the absence of school-gardens.

"This Cornell nature-study movement is one small part of a general awakening in educational circles looking toward bringing the child into actual contact and sympathy with the objects with which he has to do. This work is taking on many phases. One aspect of it is its relation to the teaching of agriculture and to the love of country life. This aspect is yet in its early experimental stage. The time will come when some institution in every State will carry on work along this line. It will be several years yet before this type of work will have reached what may be considered an established condition or before even a satisfactory body of experience shall have been attained. Out of the varied and sometimes conflicting methods and aims that are now before the public there will develop in time an institution-movement of extension agriculture teaching."

A nature-study movement alone is not sufficient to awaken and reconstruct all the agricultural interests. There should be coordinate efforts outside the schools. In order merely to suggest other lines of effort—and not to commend any particular movement—the following classification of the Cornell extension work may be made: This extension activity in agriculture is regularly and systematically reaching about 75,000 people in that State. Indirectly the work spreads to far greater numbers. Several causes have combined to produce this result, four of which are paramount. (1) The people are ready for the work: they want to learn. (2) Certain persons are ready to do the work: they want to teach. (3) The persons into whose hands the work has fallen are given freedom and autonomy: they are not restricted or hampered by those in authority. (4) The State appropriates money: the appropriation is made because work is done.

Of these four factors, the money is the least. No institution is so poor that something cannot be done if only the first three requisites are present. Time by time, perhaps little by little, the money will come. The work must be born, grow and mature. Only flies and their like are born full size.

Any good extension work is only a diligent effort to meet the needs of the people. If conditions seem to demand a certain kind of effort, that effort is made. No theory of pedagogics is concerned in it. Years hence, perhaps, it will be possible to found a theory on what shall have been accomplished.

From small beginnings the work has grown year by year. This is the most important fact in the entire movement. The work has entered fields that at first were not in sight. It has demonstrated the value of various kinds of effort, and has dropped those which seem to be of least efficiency. The Cornell extension work, as it is being prosecuted to-day [1902], may be displayed as follows:

1. EXTENSION TEACHING: Endeavoring to give a new point of view and a quickened enthusiasm to those who live in the country.

(*a*) *Nature-Study:* Teaching the youth to see and to appreciate whatever is nearest at hand, thereby bringing him into sympathy with the conditions in which he lives. This work is prosecuted by several means:

1. By reaching the rising generation. The school children in the grades are organized into Junior Naturalist Clubs to the end that they may love the country better and be content to live therein. Each club receives an embellished charter. Many thousand children are organized each year. For these children a "Junior Naturalist Monthly" is published suggesting topics for observation and study. Each child pays monthly "dues" by writing a letter or essay on some object that it has observed. The dues may be the composition required by the teacher, and it is sent to the nature-study office as it was written, without correction. Having paid its dues, the child receives a badge-button. The Junior Naturalist Club is organized under the general supervision of the teacher, but the detail of the work is carried by the Nature-study Bureau, thereby relieving the teacher of extra responsibilities. In fact, the enthusiasm and centralized interest which the Club introduces into the school lighten the burdens of the teacher.

Connected with the Junior Naturalist Enterprise is a Junior Gardener movement, to encourage specifically the growing of plants and the making of gardens. This movement is also promulgated through the schools. It now has attained great headway.

Not only is it educational wisdom to begin work with the children, but it is also one of the most efficient means of getting work done. If the children are once thoroughly interested in any enterprise, the enterprise will "go." The busiest and most obdurate man will listen to a child; so will parents. If you want to start a nature-study movement or to improve the school premises, arouse the children first.

2. By reaching the teacher directly for the purpose of reaching the pupil. For the teacher "Nature-study Leaflets" have been prepared, giving in each issue a suggestive presentation of some nature-study topic, together with notes of help and suggestion. For those teachers who desire to pursue the subjects further, a home reading course is organized and a "Home Nature-study Lesson" is published.

3. By interesting the teaching fraternity in general, through lectures at teachers' institutes and conventions, attendance on particular schools where work is being done, and other personal work. A lecturer is employed to attend State teachers' institutes, occupying a regular period on the program; this work is possible through the cooperation of the State Department of Public Instruction.

4. By summer-school teaching in the teachers' schools conducted by the State Department of Public Instruction. For two years a special nature-study summer-school was held at Cornell University, but being obliged to husband the resources this enterprise was reluctantly dropped.

5. By nature-study instruction in the University, given to those teachers who desire it.

6. By interesting the public in plant-growing, particularly in the improvement of school-grounds and the planting of gardens.

7. By direct personal correspondence with parents, teachers, ministers and other interested parties.

(*b*) *A Farmers' Reading-Course;* inducing actual farmers to pursue definite courses of reading in the winter season. The farmer who desires to read books will help himself. In this work, the effort is made to gain the attention of those who do not read books. The literature is furnished by the University, being written by members of the Extension Staff. This literature is in the form of easy eight-page "Reading-Lessons," detailing principles. Each lesson is accompanied by a set of questions, the answers of which are sent to the Bureau, entitling the reader to remain on the rolls. The Reading-Lessons are in three series of five each, as follows:

First-year series, on soil and plant-food.

Second-year series, on stock-feeding and dairying.

Third-year series, on fruit-growing.

Each reader takes these series in course. If any one desires to continue his reading beyond the third year, he is recommended to books.

The readers are aided in the formation of Reading-Clubs, to meet twice each month for the five winter months, thereby devoting two discussions to each lesson. Inspectors and lecturers visit the clubs.

The Reading-Club may arrange for experiments on local agricultural difficulties, to be conducted during the summer. This may be expected to maintain the interest throughout the busy season.

The culmination of the Reading-Course is an eleven weeks' term of instruction at the University in the winter, to which readers and others are eligible.

Reading-Course and text-book work must not be confounded with true nature-study work. The former aims directly at the imparting of information; the latter seeks to put one in sympathy with his surroundings. Any successful reading-course work brings the reader into sympathy with nature, but that is not its prime motive. The nature-study bulletin is distinct from the agriculture or farming bulletin, however elementary the latter may be.

Coordinate with the regular farmers' Reading-Course, there is a course for farmers' wives. The most difficult and discouraging feature of American agriculture is the isolated position of the farmer's wife. This position can be alleviated only by the elevation of the general tone of farm life. The farmers' wives' course is modeled after that for farmers, but it has its own literature. The publications of the Farmers' Wives' Reading-Course are thus far as follows:

Saving Steps,

Home Sanitation,

Saving Strength,

Food for the Farmer's Family,

The Kitchen Garden,

Practical Farm Housekeeping (two lessons),

Reading in the Farm Home.

[Those who desire a history of the farmers' reading-course movement should consult Bull. 72, Office of Experiment Stations, U. S. Department of Agriculture.]

2. ITINERANT EXPERIMENTING: Endeavoring to solve local agricultural perplexities by experiments on the spot, and also to illustrate the application of well-known knowledge. These experiments are of many kinds, conducted in many places. This is necessarily so, because the difficulties of farmers are so many and various. Certain definite series of illustrative experiments, have been planned from the central station, however, and farmers have been asked to cooperate. Chief of these are experiments with fertilizers, sugar beets, spraying orchards, potato and bean culture, cover-cropping, alfalfa-growing, poultry-raising. Experts are sent to investigate outbreaks of insects, fungous attacks on plants, diseases of stock, and other special difficulties. Experiments on various problems intimately associated with the extension work are also made at the University itself. Much of the results of the experimental work connected with the extension enterprise has appeared in bulletins; but its chief value is not in its publication, but in its educational effect in the communities in which it is conducted.

All this looks large and complete when seen in type, but it is the merest beginning of what should be and can be done. Other lines of effort must be added. In many places similar work is in progress. The great agricultural States of the middle West promise to become leaders. The efficiency of the work will depend in large measure on its adaptability to the particular conditions and people to be served.

The ideals of nature-study are everywhere the same; but the methods and means are capable of endless modification. There is always danger that too much emphasis will be placed on mere "learning" on the part of the child or the pupil. The real value of the extension work with the young lies in interesting, enthusing, inspiring them. Mere information, however valuable, will not cause a person to be a farmer, nor incline him to live in the country. Of course the work must be practical—that is, it must be truthful, direct, forceful, and must put the child into intimate contact with its own life. It must aim to give him power and enterprise rather than assorted facts—although the facts may be so handled that they become the means and not the end. I fear that some good persons are too insistent on getting "agriculture" into the schools. There is no gain in getting the word into the curriculum unless the subject is really taught with optimism and with purpose.

It is a common desire to bring the rural schools into intimate relations with the life of the community merely by employing teachers having knowledge of farm life. This may be of little consequence: the first merit of a teacher is to be able to teach, whatever his sympathies or technical knowledge. Many good persons seem to think that the only thing to do to reform any school problem is to get a teacher, forgetting that, in the long run, teachers arise in response to a general demand, or at least must be supported by a public sentiment. It is really beginning with the wrong end of the problem merely to ask for teachers having knowledge of agriculture. We should first awaken a general desire on the part of patrons for the new type of instruction: when this desire is aroused, the teachers will be found. Usually more can be done by beginning with the children rather than with the teacher. The children can be aroused by some outside agency. This is the meaning of the Junior Naturalist movement in New York State. Probably the true way to bring the rural school into intimate touch with rural affairs is to begin both with patrons and teachers, placing far the greater dependence on the work with patrons—and with the patrons the best results are to be expected from work with the children. By interesting the parents we shall bring pressure to bear on local school boards, school commissioners and superintendents, and school teachers to provide more usable and direct instruction.

Children are always ready to "do something." The success of kindergarten and school-garden work rests on this common trait. The school-garden idea can be variously modified. A recent adaptation of it is the "district school experiment garden" projected by O. J. Kern,[4] Superintendent of Schools of Winnebago County, Illinois. These Illinois gardens are designed for the explicit teaching of agricultural subjects. Is it not strange that schools in farming communities should not be equipped with a bit of farmed land? Aside from the tilled school-garden, why not make arrangement with the adjoining farmer to pasture his stock next the school-ground now and then? And why not have this farmer give the children talks about the animals?

In recent years there has been a marvelous application of knowledge and research to agricultural practice. We have exerted every effort to increase the productiveness and efficiency of the farm, and we have entered a new era in farming—a fact that will be more apparent in the years to come than it is now. The burden of the new agricultural teaching has been largely the augmentation of material wealth. Hand in hand with this new teaching, however, should go an awakening in the less tangible but equally powerful things of the spirit. More attractive and more comfortable farm homes, better reading, more responsive interest in the events of the world, closer touch with the common objects about him—these must be looked to before agriculture really can be revived. Appeal to greater efficiency of the farm alone cannot permanently relieve the agricultural status. This is all well illustrated in the attitude of children toward the farm. In a certain rural school in New York State of say forty-five pupils, I asked all those children that lived on farms to raise their hands: all hands but one went up. I then asked all those who wanted to live on the farm to raise their hands: only that one hand went up! Now, these children were too young to feel the appeal of more bushels of potatoes or more pounds of wool, yet they had this early formed their dislike of the farm. Some of this dislike is probably only an ill-defined desire for a mere change, such as one finds in all occupations, but I am convinced that the larger part of it was a genuine dissatisfaction with farm life. These children felt that their lot was less attractive than that of other children; I concluded that a flower garden and a pleasant yard would do more to content them with living on the farm than ten more bushels of wheat to the acre. Of course, it is the greater and better yield that will enable the farmer to supply these amenities; but at the same time it must be remembered that the increased yield itself does not awaken a desire for them. I should make farm life interesting before I make it profitable.

These points of view are well expressed by David Felmley,[5] President of the Illinois State Normal School, at Normal: "It is evident that the agricultural experiment station will never accomplish its purpose unless there is diffused among our farming population an elementary knowledge of the sciences relating to agriculture. The rural schools and the high schools attended by farmers' sons must provide the necessary instruction. There seems no other practical way. The special instruction offered in this line is not merely to train skilful farmers. It is quite important that farmer boys and girls learn to appreciate and love the country. There need be here no division in material or method. The knowledge of soil and atmosphere, of plant and animal life that makes him an intelligent producer, puts him in sympathetic touch with these activities of nature. If the farmer as he trudges down the corn rows under the June sun sees only clods, and weeds, and corn, he leads an empty and a barren life. But if he knows of the work of the moisture in air and soil, of the use of air to root and leaf, of the mysterious chemistry of the sunbeam, of the vital forces in the growing plant, of the bacteria in the soil liberating its elements of fertility; if he sees the relation of all these natural forces to his own work; if he can follow his crop to the market, to foreign lands, to the mill, to the oven and the table; if he knows of the hundreds of commercial products obtained from his corn or the animals that it fattens: he then realizes that he is no mere toiler; he is marshaling the hosts of the universe, and upon the skill of his generalship depends the life of nations."

It will be seen at once that all these new ideals are bound to result in a complete revolution of our current methods of rural school-teaching. The time cannot be very far distant when we shall have systems of common schools that are built upon the fundamental idea of serving the people in the very lives that the people are to lead. In many places there are strong protests against the old order; in other places there are distinct beginnings of the new order. The following protest is by John J. McMahan,[6] State Superintendent of Education for South Carolina: "The old-time high school prepares for the exceptional life. There is little room for Latin and Greek and fancy learning in the system of education that looks to the future lives of the great body of breadwinners and home-builders. We must abandon the pleasing delusion that all go to school with expectation of afterward going to college. We know that hardly one in a hundred will ever go to college. We define education as a preparation for complete living. Have we not adapted our preparation to the unusual and improbable life, and largely neglected preparing the average man for the duties almost certain to be upon him? We should recognize that complete living is a relative term, and that the complete life which is the ideal of the philosopher, and of the statesman as well, is not the complete life that can be realized at this stage of human development by any great number of our citizens. In holding up a high standard of education as the ultimate right of every citizen, let us not be so unmindful of the present as to deny to nearly all that education which could be given them to their great benefit and happiness."

The beginnings of the new order are seen in the nature-study movement, the establishing of agricultural high schools, the strong agitation for county or district industrial schools, the spread of reading-courses, the rise of pupils' gardens, the general awakening of rural communities. Books and methods are now made for town schools rather than for country schools; the real texts for the rural schools are just now beginning to appear, and they represent a new type of school literature. In the future, the text-book is to have relatively less influence than in the past. We have been living in a text-book and museum age. All this old method is not to be complained of. The fact that so many new subjects and propaganda are coming in shows that we are in the midst of an evolution: we are in the making of progress.

This new teaching for the farmer is a most attractive field for well-directed effort. We need more teachers for it in the colleges and normal schools and common schools. The teaching in our agricultural colleges should be seized with the missionary spirit, with the desire to send out young persons who care not so much to make professors and experimenters in the great institutions, as to give themselves to spread the gospel of nature-love and of self-respecting resourceful farming through all the colleges and all the public schools. The time is coming quickly when the college or school that wants really to reach the people must teach rural subjects from the human point of view. The real solution of the agricultural problem—which is at the same time the national problem—is to give the countryman a vital, intellectual, sympathetic, optimistic interest in his daily life. For myself, if I have any gifts, I mean to use them for the spiritualizing of agriculture.

We are on the borderland of a mighty country: we are waiting for a leader to take us to its center.

Part I, Chapter VIII

Review

This chapter was cut for the third (1909) edition, although about half of the text was rearranged, revised, and worked into the first chapter of Part II. While the revised text shows a refinement of Bailey's thought, and the chapter may have always arguably been a better stylistic fit for Part II than for Part I, some of the rhetorical force is lost in the transfer. This presents in full the brief, lyrical chapter as it appeared in the first edition.

In the increasing complexities of our lives we need nothing so much as simplicity and repose. In city or country or on the sea, nature is the surrounding condition. It is the universal environment. Since we cannot escape this condition, it were better that we have no desire to escape. It were better that we know the things, small and great, which make up this environment, and that we live with them in harmony, for all things are of kin; then shall we love and be content.

All men love nature if they but knew it. The methods and fashions of our living obscure the universal passion. The more perfect the machinery of our lives the more artificial do they become. Teaching is ever more methodical and complex. The pupil is impressed with the vastness of knowledge and the importance of research. This is well; but at some point in the school-life there should be the opening of the understanding to the simple wisdom of the fields. One's happiness depends less on what he knows than on what he feels.

There are men and women who pursue science for science's sake without thought of its relation to human lives. They are the explorers of the intellectual sphere. Immensely do they extend our horizon. They add to the store of subject-matter. They make progress possible. But these persons must always be the few. They are a professional class. Most persons desire those things which have relation to the ideals of living. To them, science as science is of little moment. They cannot pursue it. It is dry. But it may be made a means of giving them closer touch with nature. If pursued too far or in too great detail, it may repel rather than attract. What we teach as science drives many a person from nature. We must reach the people; but we can reach them only by looking from their point of view. Most persons cannot be investigators. In the school-life there must come a reaction from the too exclusive view-point of science.

In the early years we are not to teach nature as science, we are not to teach it primarily for method or for drill: we are to teach it for living and for loving—and this is nature-study. On these points I make no compromise.

The best living must always be a striving for ideals. The day of the idealist is not passed. It is here. We must not allow the phenomenal development of our material progress to obscure it. We must rise to higher ideals. We must educate the child for the life of the next generation. A good teacher has the gift of prophecy. The twentieth century is coming in with a spiritual awakening. One sign of this awakening is the outlook natureward. The growing passion for country life is a soul-movement.

More and more, in this time of books and reviews, do we need to take care that we think our own thoughts. We need to read less and to think more. We need personal, original contact with objects and events. We need to be self-poised, self-reliant. The strong man entertains himself with his own thoughts. No person should rely solely on another person for his happiness.

The power that moves the world is the power of the teacher.

FROM PART III

Inquiries: Some Practical Inquiries and Some Ways of Answering Them

Part III, originally under the title you see above, was significantly expanded from the second (1905) to the third (1909) edition, but this section was cut. It had previously appeared in Nature Portraits *(1902), appended to the end of the essay "Science for Science's Sake" (which became the second chapter of Part II in* The Nature-Study Idea *the next year), following this two-sentence transitional paragraph: "All the above remarks are meant to differentiate nature-study from science. Various questions will at once arise in the mind of the teacher." The following is as it appears in the first (1903) edition of* The Nature-Study Idea, *between the questions "But will not this nature-study be called superficial?" and "Will not this nature-study tend still further to overburden the school?"*

But do you think that this nature-study will make investigators?

That depends on what you mean by an investigator. If you mean an inquirer, then I say that nature-study will develop the trait to perfection.

If you mean one who shall discover and record new truth by means of painstaking investigation, then I answer that nature-study will not detract from such attainment. Neither does it lead directly to that end, and this is its merit. To be an investigator is to be a professionalist or specialist; and professionalists should be developed late in the school life from the few who show talent in that direction. Nature-study is for every one, and therefore is fundamental; scientific investigation is for the few, and therefore is special. If nature-study opens the sympathies natureward, it will also increase the appreciation of science. Too much are our college students taught to make their reputations as investigators. In fact, the student who goes to college or university to study usually thinks only or mostly of investigation—of his science. I wonder whether a science is not worth acquiring as a specialty for the sake of teaching it? May not reputations be made as high-class teachers of entomology or botany, even without ever publishing a bit of technical research? It would be better if the teacher were also the investigator, but there are few persons who can make happy union of the two ideals.

Reviews of *The Nature-Study Idea*

Reviews of the First (1903) Edition

In order of publication

The Outlook, vol. 74, no. 2, May 9, 1903, p. 139. HathiTrust, https://hdl.handle.net/2027/inu.32000000713992.

Beginning with the history of the nature-study movement in this country, Professor Bailey gives his views of what influence it must have upon child-life. He offers suggestions as to methods by which the child may be interested and taught, insisting that it shall not be through books, but by being placed in direct touch with nature, and by being led not only to see her facts, but to perceive her spirit. "The Growing of Plants by Children—The School Garden" is a timely chapter. He believes the nature-study idea is fundamental to the evolution of popular education, and that it is bound to have a tremendous influence in carrying a vital educational impulse to farmers, to whom accustomed methods of education are less applicable than to other people.

"Nature Study," *New York Times Saturday Review of Books and Art,*
    May 30, 1903, p. 366. TimesMachine, https://timesmachine.nytimes.
    com/timesmachine/1903/05/30/issue.html.

An interpretation of the new school movement to put the child in sympathy with nature has been written by Prof. L. H. Bailey, who fills the
Chair of Horticulture at Cornell; he is the author of the four-volume
Cyclopedia of Horticulture, and of many textbooks on botany and on
garden craft. "The Nature Study Idea," (Doubleday, Page & Co., New
York, $1 net,) is a collection of notes and essays written during the past
six years. They form all together a manual for teachers on the methods
of nature study, showing the errors to be avoided, the short cuts available, and the results to be striven for. The first part of the book deals
with "what nature study is," the meaning of the term, the origin of the
movement, its agricultural phase, and the school garden, the growing
of plants by the children. The second part treats of the "interpretation
of nature," giving extrinsic and intrinsic views of it. It has a chapter
on the efforts of the unschooled teacher to "find a use" for everything.
This begins:

> Each pupil had a plant of the Spring buttercup. The teacher called atten-
> tion to the long fibrous roots, the parted leaves, the yellow flowers, but these
> parts were apparently only incidentals, for she touched on them only lightly.
> But the hairs on the stem and leaves were important. They must be of some
> use to the plant. What is it? Evidently to protect the plant from cold, for does
> not the plant throw up its tiny stem in the very teeth of Winter? It was clear
> enough, and thus we are taught that not the least thing is made in vain.[1]

Another chapter tells of the poetic interpretation of nature, headed by
William Cullen Bryant's bob-o'-link song. The last part is entitled "Some
practical inquiries and some ways of answering them." Here are some
of the questions, each of which is answered in detail: How shall I know
what subjects to choose? How shall I make a start? How much apparatus
do I need? Will not this nature study tend still further to overburden the
school? How shall I acquire sufficient knowledge to enable me to teach
nature study? Why should this nature study be confined to schools? What
shall we do with the children in the Summer vacation? Would you advise
me to take up nature study teaching?

"Nature-study in education," *Dial*, vol. 34, no. 408, June 16, 1903, p. 405. Hathitrust, https://hdl.handle.net/2027/chi.78013595.

It is a large place in education which Professor L. H. Bailey claims for "The Nature-Study Idea" (Doubleday, Page & Co.), and a place, moreover, not held by any other subject in the school curriculum. It is not a mere adjunct to an already over-crowded course of study, but a fundamental epoch-making movement which will touch the masses with a new educational impulse and bring a stronger and more resourceful life to the pupil led by this means into a fuller and more intimate sympathy with Nature and his environment. While all readers of this stimulating and suggestive book may not be so sanguine as the author in his hope that nature-study will relieve the school-room of perfunctory methods and of desiccated science, none will fail to see the promise for great effectiveness in this direction which this new view-point brings to primary education. The thing itself, not the book about it,—the living bobolink, not even the stuffed specimen,—the process of discovery, rather than the fact observed,—these stamp the nature-study idea as revolutionary in educational methods. It is not science, but a method which has room for fancy and sentiment as well as fact, and its net result is a little knowledge and more love of Nature's forms and an independent habit of seeing things intelligently as they really are. In this lies the solution of the agricultural problem, the spiritualizing of agriculture, and also the ground for a new ethics of sport with gun and rod and of man's relations to other living things. Seekers for definite schedules of courses, specific directions for nature-study lessons, or illustrations of matter and method, will be disappointed in Professor Bailey's treatise; but those who seek inspiration will find his pages breathing that spirit which gives life in all things.

J. E. Davis, *Southern Workman*, vol. 32, no. 7, July 1903, p. 342. HathiTrust, https://hdl.handle.net/2027/uc1.b3455587.

The firm of Doubleday, Page and Co., New York, has recently published three books of special interest to public-school teachers. They are "The Nature-Study Idea," by L. H. Bailey; "How to Make School Gardens," by H. D. Hemenway; and "More Money for the Public Schools," by Charles W Eliot. Price $1.00 each.

The Nature-Study Idea is announced by its author, the head of the Agricultural Department of Cornell University and the founder of the Nature-Study Bureau of that institution, to be an interpretation of the new school movement to put the child in sympathy with nature. "Nature study," he says, "is not science. It is not knowledge. It is not facts. It is spirit. It is concerned with the child's outlook on the world." It is because of the common misconception of the meaning and mission of the nature-study movement that the book has been prepared. Part I is devoted to a detailed statement of what, in Dr. Bailey's opinion, constitutes nature study as distinguished from elementary science. Part II treats of the various methods of interpreting nature, and Part III concerns itself with replies to practical inquiries and objections in regard to the introduction of this subject into the common schools.

After treating briefly of the origin of the name "nature study," which he feels we would do better to call "nature sympathy," Dr. Bailey considers the real meaning of the term as he understands it, and explains how he thinks it should be taught. The results of teaching it, he says, should be the developing of mental power, the opening of the eyes and the mind, the civilizing of the individual. "Nature study," he declares, "not only educates, but it educates nature-ward; nature-love tends toward naturalness, and toward simplicity of living. It tends country-ward." He feels that the keynote of such teaching is sympathy and that it means such a presentation to children of the outside world that they will learn to love all of nature's forms and cease to abuse them. The *spirit* of nature study, Dr. Bailey thinks, will survive in the schools and give them a new impulse, even though the name should disappear.

The chapter on school gardens is most helpful. The fact is recognized that the public conscience must first be appealed to in order to improve school grounds; then, before the school garden, must come the cleaning up and planting of the yard. The author's advice is to take time and not expect too much at once. Teachers can hardly fail to get inspiration from this little book, together with much practical and useful information in regard to ways and means. The section devoted to the answering of questions concerning the teaching of nature study that have occurred to many a teacher is of special value. [. . .]

Dallas Lore Sharp, from "Nature between Book Covers," *Critic*, vol. 43, no. 2, August 1903, pp. 166–168. HathiTrust, https://hdl.handle. net/2027/iau.31858055204980?urlappend=%3Bseq=107%3Bowne rid=13510798903717309-111.

This bundle of nature books tumbled from the package like a lapful of flowers, gay of cover and summery. You cannot think of a nature book without a cover, a distinctive cover; but nature books with only covers—and photogravures—are as common these days as daisies. A few photographs, a little of Packard's text, a gorgeous Polyphemus cover—and you have the latest how-to-know moth book. Photographs, bits of Gray's text, a landscape cover in gold—a how-to-know tree book. The how-to-know bird books are like the moth and tree books—only more than both of them for number.

I would not trade my businesslike Chapman for all the colored, becameraëd bird books of the last decade, nor my little leather-covered Gray for all the recent botanical books I have seen.

Of course Gray is not "popular"—it has no pictures, does not announce on the title-page "For non-botanical readers," nor commence with poetry. But then, everybody goes a-naturing now, and doubtless these pretty books have helped start the crowd off. [. . .]

Let us hear the conclusion of the whole matter [of nature-study], says Professor Liberty H. Bailey in his book of essays, "The Nature-Study Idea." The whole matter of nature study is not concluded by Professor Bailey, but he answers a big number of questions that have bothered us, and he sets going a very much bigger number of new ones that it will do us good to ponder on. What is nature study? Its dangers? Its benefits? What are we in relation to the world about us? What is worth while anyhow? You may not agree with the little book, but you will think, and the outdoor world will seem a very good and wholesome place to live in after you have read "The Nature-Study Idea." The scoffers will still scoff at the beans in the schoolroom windows. There won't be so many scoffers after this, however, and there will be more beans. There is little excuse for the slovenly style and for the mistakes in this book. The author says, for instance: "Botany has to do with cells and protoplasm and cryptoGRAMS."[2]

Journal of Education, vol. 58, no. 12, September 24, 1903, p. 215. SAGE Journals, https://journals-sagepub-com.proxy.library.nyu.edu/toc/jexa/58/12.

Professor L. H. Bailey of Cornell University is by far the best equipped by taste, with information, by experience, in culture, and by platform gifts to be the leader, the master, and the genius of the nature study movement. He is the great leader of the professional horticulturists, floriculturists, and agriculturists, and is withal an eminently satisfactory leader and teacher of teachers.

This little manual should be in the hands of every teacher of nature study, and no teacher has any right not to teach this subject. Mr. Bailey robs it of all the nonsense, weakness, and rubbish which have been associated with the subject. He tells in eight short chapters what nature study is, then in seven chapters he treats of the interpretation of nature, and finally answers twenty-six practical, everyday inquiries in a thoroughly sensible manner.

Journal of Education, vol. 58, no. 23, December 10, 1903, p. 407. SAGE Journals, https://journals-sagepub-com.proxy.library.nyu.edu/toc/jexa/58/23.

And still the wonder grows that any man can find time to know so much about all sides of nature life, to be so accurate in all his observations, so complete in all the detailed information, and so uniformly entertaining, without sacrificing fact or dealing in fancy.

This interesting volume is an illuminating and suggestive study of the new movement, originating in the common schools, to put the child into sympathy with nature and his environment, to the end that his life may be stronger and more resourceful. This movement relates education directly to the life that the pupil is to live. It is a fundamental, epoch-making movement. It is a revolt from mere science-teaching in the grades and from all perfunctoriness in school work. It is the full expression of personality. It is not the mere addition of certain studies to a curriculum, but the inspiration of a new point of view in education. More than any other recent movement, it will touch the masses with a new educational impulse.

REVIEWS OF THE THIRD (1909) EDITION

In order of publication

Annals of the American Academy of Political and Social Science, vol. 35, no. 1, January 1910, pp. 185–186. SAGE Journals, https://journals-sagepub-com.proxy.library.nyu.edu/toc/anna/35/1.

The thesis of this book—not stated, but read on every page—is that elementary education should consist in adjusting the student to the world. "The happiest life has the greatest number of points of contact with the world," and Nature Study is the most natural and forceful way of multiplying these points of contact for the child. The force of many chapters is devoted to making clear the distinction between Nature Study and Science, for it is a confusion of these terms that breeds the chief opposition which technical scientists hold for the subject. Science gives information—Nature Study gives spirit; Science is of the intellect—Nature Study is of the heart. A teacher who thinks first of his *subject* teaches science; one who thinks first of his *pupils* teaches nature study. The two cannot conflict, for they occupy different fields. Part II of the book, entitled "The Teacher's

Outlook to Nature," is a series of rather unrelated papers on the interpretation of nature. The volume closes with replies to miscellaneous queries propounded to the author concerning the teaching and advancement of nature study.

Cornell Countryman, vol. 7, no. 8, May 1910, p. 278. HathiTrust, https://hdl.handle.net/2027/coo.31924093392292.

If a book merely presents a new idea or food for some original thought, it is worthy of note. If that book has combined with the new idea a beautiful flow of language, the outpourings of the soul of a naturalist, and the thoughts of a great genius, it is certain that that book will live. Such a one is Dean Bailey's new work.

Without doubt, the majority of people who read "The Nature-Study Idea" will disagree with the author and criticize him for expressing wild ideas. It is very easy to understand why this will happen. This book presents an entirely different conception of Nature-Study from that in common vogue today. It destroys its position as a science and tears down the fine technical framework, which grammar school teachers, high-school instructors, and college professors have constructed around this study. It destroys the erratic idea that everything must have a use. It gives a chance for each person to develop from early childhood to the spirit of a naturalist. It will open the eyes of each reader to the beauties—the concealed beauties—of Nature and to the wonderful personality of the author.

The last part of the book is devoted to questions and answers. In this part, every possible objection is met. Read and be convinced.

Related Writings

Note on the Selections

The texts in this section appear in roughly chronological order, beginning at the start of the Cornell Nature-Study Bureau's publishing efforts in 1896, through the initial drafting of *The Nature-Study Idea* around 1901, its publication in 1903, and its major revision in 1909, and into the years of Bailey's early retirement, culminating with two addresses on education given in 1918. The aim has been to present both some of the texts centrally referred to in *The Nature-Study Idea* and also supplementary texts that demonstrate the ways Bailey continued to develop and refine the book's ideas across some of his most productive literary years. In addition to further developing the themes of *The Nature-Study Idea*, these texts begin to illustrate the ways in which nature-study formed the core of his larger country life philosophy and theory of change.

The selection begins with "How a Squash Plant Gets Out of the Seed," the first of Cornell's series of Teachers' Leaflets, which would become the model for all the future leaflets published by the College of Agriculture and written generally by its faculty and nature-study staff. This is the pamphlet that Bailey defends against "the Integument-Man" in his chapter of that title in *The Nature-Study Idea*, the scientist who accuses Bailey of misleading children by using the word "seed" instead of "integument." The defense that Bailey lays out of his use of the more common word "seed," which he knew children would recognize as the appropriate general word for what a squash plant comes out of, would be used throughout Cornell's publication efforts to defend the use of common language to describe phenomena common to the child's life. The goal was not to teach specialist terminology to young children but to lead them to a deeper general understanding of and sympathy with the natural world. And despite terminological quibbles, the leaflet provides remarkably detailed analysis, translated into the language of the child. These leaflets were intended for teachers (a separate series was later started for children's direct use), but they were written to engage teachers and model an effective way of speaking to students, and they were richly illustrated to suggest to the teacher an engaging way of viewing the physical material to be used in class. The leaflet was meant to guide and supplement direct, experiential nature-study work but not to replace that work. It was frequently reprinted, sometimes with minor revisions. The text and images presented here are as they appeared in the very first edition of 1896; interestingly, the term "nature-study" had not yet come to replace "natural science" in this edition, although it would in the leaflet's subsequent printings.

That pamphlet is followed by the series of articles that can be said to have provided the impetus for *The Nature-Study Idea*'s publication (more on this in the essay "It Is Spirit," this volume). The first of these is another of Bailey's contributions to Cornell's nature-study leaflet series for teachers in the public schools. After the squash leaflet and four more initial leaflets that each explored different specific subjects recommended for nature-study lessons, this issue sought to present the kernel of the nature-study philosophy adopted by Bailey and the team at Cornell under the simple title "What Is Nature-Study?" The text is that of the "second edition" published in 1897, which is the earliest I have been able to locate. (It seems that "editions" referred to printings, as a way to measure the demand for these state-funded leaflets, regardless of whether alterations were made or any text was actually reset.) Following this short manifesto, I have presented two essays written by Bailey's undergraduate professor and mentor, William James Beal, as they first appeared in the

"Correspondence" section of the journal *Science*, each of which critique Bailey's philosophy as lacking rigor and inviting scientific inaccuracy and "sentimentality." Beal pointedly uses Bailey's title for both of his short essays, and he quotes Bailey's original "What Is Nature-Study?" essay as representative of such "injurious" nature-study work. Just four months after the second article was published in *Science*, *The Nature-Study Idea* was released by Doubleday, and it addressed Beal's second essay on the very first page (albeit not by name), definitively reclaiming the title "What Is Nature-Study?" for the book's first chapter. The whole book can be read as a response to Beal's criticism and that of the "eminent scientific men" (and one woman) whom Beal cites. For the full effect, after reading Beal's two letters to *Science*, the reader might revisit the first pages of the first chapter of *The Nature-Study Idea*.

Cornell's various series of nature-study leaflets were so popular and so frequently reprinted to meet the demand for them that a large selection was revised and edited into book form in 1904 under the simple title *Cornell Nature-Study Leaflets*. The fourth and final piece reproduced in the "What Is Nature-Study?" series here, following Beal's essays, is a revision and enlargement of Bailey's initial leaflet of that title, this time as it appeared in the 1904 compilation volume, one year after *The Nature-Study Idea* appeared. In this new context, it was once again used as the volume's opening essay, following Bailey's poem "The School House" (revised and reprinted in other contexts under the name "The Country School" or "Country School"), and it was followed by several more of his essays—"The Nature-Study Movement," "An Appeal to the Teachers of New York State," and "What Is Agricultural Education?"—forming a forty-five-page treatise by Bailey at the beginning of the volume. The remainder of that 607-page tome included leaflets (most of them lavishly illustrated) by numerous of Cornell's nature-study faculty, including Anna Botsford Comstock, Mary Rogers Miller, Alice G. McCloskey, John W. Spencer, and others, as well as many more by Bailey himself. Edward F. Bigelow, who was then the editor of *Nature and Science*, wrote to Bailey that the compilation volume "literally FILLS A LONG-FELT WANT." It was reprinted by J. B. Lyon Company, and demand was so high that by 1906 the New York Commissioner of Agriculture wrote to Bailey that the state had exhausted their supply and were requesting additional copies.[1]

Cornell Nature-Study Leaflets also set the model for Anna Botsford Comstock's later *Handbook of Nature Study*, which similarly brought together previously published leaflets but more fully integrated them into a whole. Comstock's *Handbook* would become a best-seller and remains in print and in demand today, so Bailey's opening essay from the earlier volume holds historic interest for its place in that publication lineage as well.

Following this exchange on the definition and scope of the nature-study idea is an essay from 1901, published in *The American Monthly Review of Reviews*, in which Bailey more explicitly describes the exact work in which the Cornell nature-study program was engaged at about the time that he would have been composing the first draft of *The Nature-Study Idea*. "Nature-Study on the Cornell Plan" lays out the program more clearly than the book does and in a way that apparently was meant to be replicable by other public land-grant colleges. It also is of special interest for its description of the Junior Naturalist Club program. Bailey once again stresses the work's importance to rural society, although he notes that it was even more quickly embraced by urban schoolteachers and that a correspondence program linking rural and urban students worked to help bridge the rural/urban divide. The same issue of the *Review of Reviews* featured an essay about a movement in Michigan to foster cooperation and sympathy between rural teachers and the parents of schoolchildren, written by Kenyon L. Butterfield, who would later serve alongside Bailey on Theodore Roosevelt's Commission on Country Life.

The next essay presented here is "The Common Schools and the Farm-Youth," which was published in *The Century Magazine* (the successor to *Scribner's Monthly Magazine*) in 1907, just two years before the third, revised edition of *The Nature-Study Idea* appeared. Bailey notes that the essay concludes a series of three previous articles of his that had appeared in the magazine the year before. The series focused on the relationship of young people toward the country, from his position as director of the College of Agriculture at Cornell University (a position he had held since shortly after the first edition of *The Nature-Study Idea* was published in 1903). It therefore continues to explore the place of his nature-study philosophy in the larger country life movement. It also represents Bailey's thoughts the year that *The State and the Farmer* appeared, when he would have been pulling together The Rural Outlook Set and shortly before he became chair of the Commission on Country Life. In the essay he continues to center the public schools as leaders in the nature-study movement even as he bemoans the fact that there weren't more schools embracing the work. As with Bailey's later country-life work, in this essay he emphasizes nature-study's grassroots origins while de-emphasizing his own contributions to the movement, framing the essay as a discussion of the nature-study publications that had appeared across the country in the years since the movement began.

"The Common Schools and the Farm-Youth" is of further interest for including a floorplan and sketches of the model rural schoolhouse built at Cornell for nature-study work the very year that the article appeared. The two-room schoolhouse was completed in 1907, the same year the essay was published, alongside several major academic buildings begun in 1905. The groundbreaking for this major expansion of the college's infrastructure featured Bailey guiding a plow pulled by the college's agri-culture students. Bailey had been planning the small model schoolhouse at least since 1903, however, and historian Gould P. Colman argues that the schoolhouse was, "[m]easured by Bailey's personal interest, the most important of the new buildings," despite being among the most humble in size.[2] The school's workroom, where hands-on nature-study could be pur-sued, was the smaller of the two rooms, but Bailey believed that it would have to be enlarged as the nature-study work developed. Unfortunately, when the building was completed, Cornell's trustees refused to let Bailey organize it as a working school along the lines of John Dewey's laboratory school in Chicago, and it was instead leased for use as a private school. It was demolished in 1962, but for many years it stood directly in front of Bailey Hall, a large auditorium and lecture hall built and named in Bailey's honor after his retirement, and the humble schoolhouse represented some-thing of Bailey's highest hopes for rural society.[3]

Rounding out these selections are three essays published about a decade later. After the stressful years of national leadership brought on by his role on Roosevelt's commission, Bailey stepped away from admin-istrative work, retiring at an unexpectedly early age from the deanship at Cornell to devote his time to writing and editing the philosophical as well as scientific projects he had been putting off for years, and among the subjects he returned to was nature-study. First of these is a short personal sketch from the pages of *The Nature-Study Review* of 1916, "When the Birds Nested." By this time, Bailey had retired from his formal work at

Cornell and published the first two volumes of his Background Books series, his philosophical manifesto, *The Holy Earth* (1915), and collection of poetry, *Wind and Weather* (1916). Anna Botsford Comstock had taken over editorship of the *Review* and was featuring in it more lyrical pieces like Bailey's short essay, which channels the kind of personal writing found in Part II of *The Nature-Study Idea*. It also provides further insight into how Bailey thought about his own childhood in relation to the ideals of the nature-study movement, and it is notable for its depiction of the massive flocks of passenger pigeons he remembered from those years (cf. II.V). He contributed several new poems, as well as other short prose pieces, to the *Review* in this period.

    The final two selections are a pair of addresses from 1918, titled "The Science Element in Education" and "The Humanistic Element in Education." Bailey would have given these in the same year that he published the third and fourth Background Books, *Universal Service* and *What Is Democracy?*, in which he laid out his agrarian theory of democracy at a pivotal moment as the United States entered the first World War. The role of education to the development of democracy was weighing heavily on his mind. The main thrust of both addresses—the first presented to the Central Association of Science and Mathematics Teachers and published in *School Science and Mathematics*, and the second presented as the president's address at the annual meeting of the American Nature-Study Society and published in *The Nature-Study Review*—centers on the arbitrariness of disciplinary boundaries and a call to break down the artificial wall separating the sciences and the humanities, similar to the call he had made many decades earlier to break down the "garden fence" separating horticulture from botany.[4] The full text of his address on "The Science Element in Education" was not reproduced, but a generous selection appeared in the educational journal, and the "abstract" of the remainder of the essay (also reproduced here as it appeared) seems to channel the language Bailey would have used in the speech, if it wasn't in fact written by Bailey himself in the third person. "The Humanistic Element in Education" appears to have been reproduced in full for *The Nature-Study Review*, and it appears here as it did there. Together, the two addresses consider the enterprise of education broadly in its value to the cultivation of the full individual and as a contributor to democracy, and they illustrate the ways in which much of Bailey's larger philosophical vision emanated from the idea and practice of nature-study.

Teacher's Leaflets for Use in the Rural Schools.[*]
Prepared by the Agricultural Experiment Station of Cornell
University, Ithaca, N. Y.
Issued under the auspices of the Experiment Station
Extension, or Nixon Law.
By L. H. Bailey.
No. 1, Dec. 1, 1896.

How a Squash Plant Gets Out of the Seed.

---

[*] NOTE.—These leaflets are intended for the teacher, not for the scholars. It is their purpose to suggest the method which a teacher may pursue in instructing children at odd times in nature-study. The teacher should show the children the objects themselves,—should plant the seeds, raise the plants, collect the insects, etc.; or, better, he should interest the children to collect the objects. Advanced pupils, however, may be given the leaflets and asked to perform the experiments or make the observations which are suggested. The scholars themselves should be taught to do the work and to arrive at independent conclusions. Teachers who desire to inform themselves more fully upon the motives for this nature-study teaching, should write for a copy of Bulletin 122, of the Cornell Experiment Station, Ithaca, N. Y.

How a Squash Plant Gets Out of the Seed.

By L. H. Bailey.

If one were to plant seeds of a Hubbard or Boston Marrow squash in loose warm earth in a pan or box, and were then to leave the parcel for a week or ten days, he would find, upon his return, a colony of plants like that shown in Fig. 11. If he had not planted the seeds himself or had not seen such plants before, he would not believe that these curious plants would ever grow into squash vines, so different are they from the vines which we know in the garden. This, itself, is a most curious fact,—this wonderful difference between the first and the later stages of all plants, and it is only because we know it so well that we do not wonder at it.

It may happen, however,—as it did in a pan of seed that I sowed a few days ago—that one or two of the plants may look like that shown in Fig. 12. Here the seed seems to have come up on top of the plant, and one is reminded of the curious way in which beans come up on the stalk of the young plant. If we were to study the matter, however,—as we may do at a future time,—we should find a great difference in the ways in which the squashes and the beans raise their seeds out of the ground. It is not our

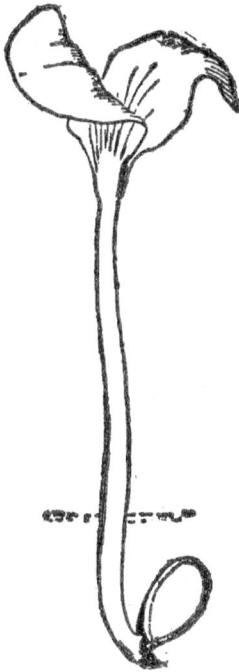

Figure 11. Squash plant a week old.

Figure 12. Squash plant which has brought the seed-coats out of the ground.

purpose to compare the squash and the bean at this time, but we are curious to know why one of these squash plants brings its seed up out of the ground whilst all the others do not. In order to find out why it is, we must ask the plant, and this asking is what we call an experiment.

We may first pull up the two plants. The first one (Fig. 11) will be seen to have the seed-coats still attached to the very lowest part of the stalk below the soil, but the other plant has no seed at that point. We will now plant more seeds, a dozen or more of them, so that we shall have enough to examine two or three times a day for several days. A day or two after the seeds are planted, we shall find a little point or root-like portion breaking out of the sharp end of the seed, as shown in Fig. 13. A day later this root portion has grown to be as long as the seed itself (Fig. 14), and it has turned directly downwards into the soil. But there is another most curious thing about this germinating seed. Just where the root is breaking out of

Figure 13. Germination just beginning.

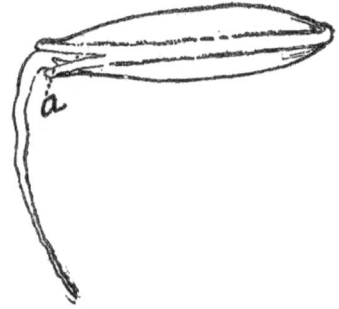

Figure 14. The root and peg.

the seed (shown at *a* in Fig. 14), there is a little peg or projection. In Fig. 15, about a day later, the root has grown still longer, and this peg seems to be forcing the seed apart. In Fig. 16, however, it will be seen that the seed is really being forced apart by the stem or stalk above the peg for this stem is now growing longer. The lower lobe of the seed has attached to the peg (seen at *a*, Fig. 16), and the seed-leaves are trying to back out of the seed. Fig. 17, shows the seed still a day later. The root has now produced many branches and has thoroughly established itself in the soil. The top is also growing rapidly and is still backing out of the seed, and the seed-coats are still firmly held by the obstinate peg.

Whilst we have been seeing all these curious things in the seeds which we have dug up, the plantlets which we have not disturbed have been coming through the soil. If we were to see the plant in Fig. 17, as it was "coming up," it would look like Fig. 18. It is tugging away trying to get its head out of the bonnet which is pegged down underneath the soil, and it has "got its back up" in the operation. In Fig. 19, it has escaped from its trap and is laughing and growing in delight. It must now straighten itself up, as it is doing in Fig. 20, and it is soon standing proud and straight, as in Fig. 11. We now see that the reason why the seed came up on the plant in Fig. 12, is because in some way the peg did not hold the seed-coats down (see Fig. 23), and the expanding leaves are pinched together, and they must get themselves loose as best they can.

There is another thing about this curious squash plant which we must not fail to notice, and this is the fact that these first two leaves of the plant-let came out of the seed and did not grow out of the plant itself. We must

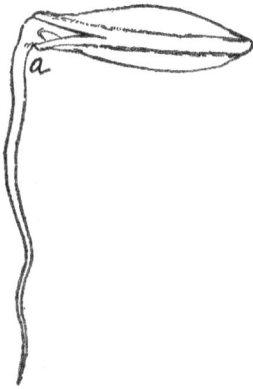

Figure 15. Third day of root growth.

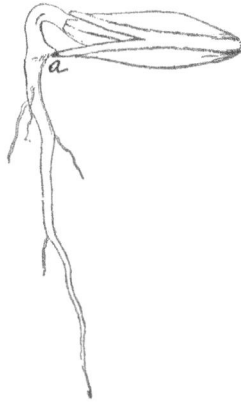

Figure 16. The plant breaking out of the seed-coats.

Figure 17. The operation further progressed.

Figure 18. The plant just coming up.

Figure 19. The plant liberated from the seed-coats.

Figure 20. The plant straightening up.

notice, too, that these leaves are much smaller when they are first drawn out of the seed than they are when the plantlet has straightened itself up. That is, these leaves increase very much in size after they reach the light and air. The roots of the plantlet are now established in the soil and are taking in food which enables the plant to grow. The next leaves which appear will be very different from these first or seed leaves.

Figure 21. The true leaves developing.

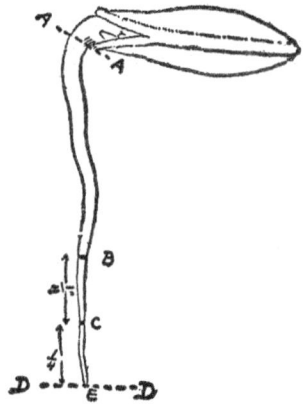

Figure 22. Marking the root.

These later ones are called the true leaves. They grow right out of the little plant itself. Fig. 21 shows these true leaves as they appear on a young Crookneck squash plant, and the plant now begins to look much like a squash vine.

We are now curious to know how the stem grows when it backs out of the seeds and pulls the little seed-leaves with it, and how the root grows downwards into the soil. Now let us pull up another seed when it has sent a single root about two inches deep into the earth. We will wash it very carefully and lay it upon a piece of paper. Then we will lay a ruler alongside of it, and make an ink mark one-quarter of an inch from the tip, and two or three other marks at equal distances above (Fig. 22).* We will now carefully replant the seed. Two days later we will dig it up, when we shall most likely find a condition something like that in Fig. 23. It will be seen that the marks E, C, B, are practically the same distance apart as before

---

* NOTE.—Common ink will not answer for this purpose because it "runs" when the root is wet, but indelible ink, used for marking linen or for drawing, should be used. It should also be said that the root of the common pumpkin, and of the summer bush squashes, is too fibrous and branchy for this test. It should be stated, also, that the root does not grow at its very tip, but chiefly in a narrow zone just back of the tip; but the determination of this point is rather too difficult for the beginner, and, moreover, it is foreign to the purpose of this tract.

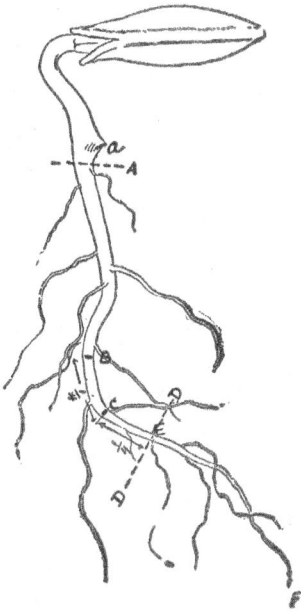

Figure 23. The root grows in the
end portions.

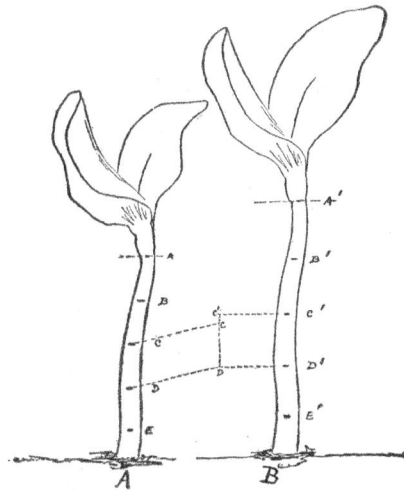

Figure 24. The marking of the stem, and the
spreading apart of the marks.

and they are also the same distance from the peg AA. The point of the root is no longer at DD, however, but has grown on to F. The root, therefore, has grown almost wholly in the end portion.

Now let us make a similar experiment with the stem or stalk. We will mark a young stem, as at A in Fig. 24; but the next day we shall find that these marks are farther apart than when we made them (B, Fig. 24). The marks have all raised themselves above the ground as the plant has grown. The stem, therefore, has grown between the joints rather than from the tip. The stem usually grows most rapidly, at any given time, at the upper or younger portion of the joint (or internode); and the joint soon reaches the limit of its growth and becomes stationary, and a new one grows out above it.

Natural science consists in two things,—seeing what you look at, and drawing proper conclusions from what you see.

Teacher's Leaflets for Use in the Public Schools
Prepared by the College of Agriculture, Cornell University, Ithaca, N. Y.
Issued under Chapter 128 of the Laws of 1897.
I. P. Roberts, Director.
Second Edition, No. 6, June 1, 1897.

WHAT IS NATURE-STUDY?

BY L. H. BAILEY

It is the seeing of things which one looks at, and the drawing of proper conclusions from what one sees. Nature-study is not the study of a science, as of botany, entomology, geology, and the like. That is, it takes the things at hand, and endeavors to understand them, without reference to the systematic order or relationship of the objects. It is wholly informal and unsystematic, the same as the objects are which one sees. It is entirely divorced from definitions, or from explanations in books. It is therefore supremely natural. It simply trains the eye and the mind to see and to comprehend the common things of life; and the result is not directly the acquirement of science but the establishment of a living sympathy with everything that is.

The proper objects of nature-study are the things which one oftenest meets. To-day it is a stone; to-morrow it is a twig, a bird, an insect, a leaf, a flower. The child, or even the high school pupil, is first interested in things which do not need to be analyzed or changed into unusual forms or problems. Therefore, problems of chemistry and of physics are for the most part unsuited to early lessons in nature-study. Moving things, as birds, insects and mammals, interest children most and therefore seem to be the proper subjects for nature-study; but it is often difficult to secure specimens when wanted, especially in liberal quantity, and still more difficult to see the objects in perfectly natural conditions. Plants are more easily had, and are therefore more practicable for the purpose, although animals and minerals should by no means be excluded.

If the objects to be studied are informal, the methods of teaching should be, also. If nature-study were made a stated part of the curriculum, its purpose would be defeated. The chiefest difficulty with our present school methods is the necessary formality of the courses and the hours. Tasks are set, and tasks are always hard. The only way to teach nature-study is, with no course laid out, to bring in whatever object may be at hand and to set the pupils to looking at it. The pupils do the work,—they see the thing and explain its structure and its meaning. The exercise should not be long,—not to exceed fifteen minutes at any time, and, above all things, the pupil should never look upon it as a recitation, and there should never be an examination. It should come as a rest exercise, whenever the pupils become listless. Ten minutes a day, for one term, of a short, sharp and spicy observation upon plants, for example, is worth more than a whole text-book of botany.

The teacher should studiously avoid definitions, and the setting of patterns. The old idea of the model flower is a pernicious one, because it really does not exist in nature. The model flower, the complete leaf, and the like, are inferences, and pupils should always begin with things and not with ideas. In other words, the ideas should be suggested by the things, and not the things by the ideas. "Here is a drawing of a model flower," the old method says; "go and find the nearest approach to it." "Go and find me a flower," is the true method, "and let us see what it is."

Every child, and every grown person too, for that matter, is interested in nature-study, for it is the natural method of acquiring knowledge. The only difficulty lies in the teaching, for very few teachers have had any drill or experience in this informal method of drawing out the observing and reasoning powers of the pupil wholly without the use of text-books. The teacher must first of all feel the living interest in natural objects which it is desired the pupils shall acquire. If the enthusiasm is not catching, better let such teaching alone.

All this means that the teacher will need helps. He will need to inform himself before he attempts to inform the pupil. It is not necessary that he become a scientist in order to do this. He simply goes as far as he knows and then says to the pupils that he cannot answer the questions which he cannot. This at once raises the pupil's estimation of him, for the pupil is convinced of his truthfulness, and is made to feel—but how seldom is the sensation!—that knowledge is not the peculiar property of the teacher but is the right of anyone who seeks it. It sets the pupil investigating for himself. The teacher never needs to apologize for nature. He is teaching only because he is an older and more experienced pupil than his pupil is. This is just the spirit of the teacher in the universities to-day. The best teacher is the one whose pupils farthest outrun him.

In order to help the teacher in the rural schools of New York, we have conceived of a series of leaflets explaining how the common objects can be made interesting to children. Whilst these are intended for the teacher, there is no harm in giving them to the pupil; but the leaflets should never be used as texts to make recitations from. Now and then, take the children for a ramble in the woods or field, or go to the brook or lake. Call their attention to the interesting things you meet—whether you yourself understand them or not—in order to teach them to see and to find some point of sympathy; for everyone of them will some day need the solace and the rest which this nature-love can give them. It is not the mere information which is valuable; that may be had by asking someone wiser than they, but the inquiring and sympathetic spirit is one's own.

The pupils will find their lessons easier to acquire for this respite of ten minutes with a leaf or an insect, and the school-going will come to be less perfunctory. If you must teach drawing, set the picture in a leaflet before the pupils for study, and then substitute the object. If you must teach composition, let the pupils write upon what they have seen. After a time, give ten minutes now and then to asking the children what they saw on their way to school.

Now, why is the College of Agriculture of Cornell University interesting itself in this work? It is trying to help the farmer, and it is beginning with the most teachable point,—the child. The district school cannot teach agriculture any more than it can teach law or engineering or any other profession or trade, but it can interest the child in nature and in rural problems and thereby fasten its sympathies to the country. The child will teach the parent. The coming generation will see the result. In the interest of humanity and country, we ask for help.

To the Teacher:

The following leaflets have been issued to aid teachers in the public schools in presenting nature-study subjects to the scholars at odd times.

1. *How a squash plant gets out of the seed.*
2. *How a candle burns.*
3. *Four apple twigs.*
4. *A children's garden.*
5. *Some tent-makers.*
6. *What is nature-study?*

Address,
     Chief Clerk,
         College of Agriculture,
             Ithaca, N. Y.

[From Science, *vol. 15, no. 390, June 20, 1902, pp. 991–992.]*

WHAT IS NATURE STUDY?

[BY WILLIAM JAMES BEAL]

THERE seem to be many conflicting definitions in attempts to answer the above question. Here are two examples: "Nature study, as used in this paper, is understood to be the work in elementary science taught below the high school—in botany, zoology, physics, chemistry and geology. We should aim to define results. Gushing sentimentalism or mere rambling talks will be as barren in results as undigested statistics. To avoid this, the teacher should always have a definite plan before her when the lesson begins."—D. Lange, Supervisor of Nature Study, St. Paul, Minn.

"Nature Study is seeing the things which one looks at, and the drawing of proper conclusions from what one sees. Nature study is not the study of a science, as of botany, entomology, geology and the like. It is wholly informal and unsystematic, the same as the objects are which one sees. It is entirely divorced from definitions, or from explanations in books. * * * To-day it is a stone; to-morrow it is a twig, a bird, an insect, a leaf, a flower. * * * The problems of chemistry and of physics are for the most part unsuited to early lessons in nature study.

"If nature study were made a stated part of a curriculum, its purpose would be defeated."—L. H. Bailey, Cornell University, N. Y.

I have observed the different methods of teaching botany and zoology for many years past. So far as this country is concerned, I think what is now correctly termed nature study started with Louis Agassiz at Harvard, where he invariably set his special students in zoology to work on a star-fish, a lobster, a clam or some other animal; not one specimen of one of these, but many of them, not alone those that were full grown, but those of all ages; not only dead specimens, but those that were alive, always with numerous comparisons. For months, the use of books was positively forbidden; and all that was told the student, excepting a few names of parts, was, 'You are right,' or 'You are wrong,' and if wrong, the student was kept at the work until he saw the thing right.

Agassiz was overflowing with enthusiasm. He would throw up both arms with exclamations of delight on seeing a specimen of a common shell-fish that was overgrown. This earnestness and enthusiasm helped secure faithful work from his students. Since working under Agassiz I have not had the slightest doubt that his method of studying nature or nature study was unsurpassed for advanced students. This method made a lasting impression on Harvard, on her presidents, her professors, and all the students who took his kind of work. Through these students of Agassiz and their students down to the third generation, this spirit of inde-pendent work has come filtering along for fifty years or more, till it has finally become widespread and deeply seated, and has recently burst forth into a great flame.

After the manner of Agassiz with his post-graduates, so the teacher of the grades below the high school will treat her young students, of course giving easier problems requiring but a little time each day. The teacher will show her interest, tact and enthusiasm to draw out the best work from her pupils. By all devices, she will seek to get the results of the combined observations of all members of her class before she lets them know her own views on the subject, and even then parts of the work may be left with pupils for further investigation.

With much that is good in nature study comes much that is positively injurious, and unfortunately large numbers are unable to distinguish between the true and the false. One writes a little book giving it some fancy title, distorts the drawings of some seeds and seedlings, inserting outlines of children's faces thereon; she writes some marvelous stories, and all these to help arouse and retain the interest of the child.

I have in my possession a neat drawing made by a student. He made two drawings to represent two honey bees just about to visit apple blossoms. The bees are not alike; each has two wings only; the heads and legs are unlike anything ever attached to bees. The apple blossoms are five-lobed (gamopetalous), with three stamens growing from the base of each lobe of the corolla. He has made drawings of imaginary insects seeking imaginary nectar from imaginary flowers. This student was trained in a state normal school. Such caricatures are absolutely worthless, in fact injurious, to any young person who makes them or even looks at them.

W. J. Beal.

Agricultural College, Mich.

[From Science, *vol. 16, no. 414, December 5, 1902, pp. 991–992.]*

WHAT IS NATURE STUDY?

[BY WILLIAM JAMES BEAL]

As was stated in SCIENCE for June 20, of this year, there seem to be, among educators, many conflicting definitions in the attempt to answer the above question. Bearing on this subject the following letters have been received from eminent scientific men of this country. They appear in the order in which they were received.

W. J. BEAL.

AGRICULTURAL COLLEGE, MICH.

The present movement toward developing and spreading an interest in nature studies is one of prime importance. Our American children are, after all the efforts thus far made, woefully lacking in interest in natural history—far behind German, and even English children, I fancy.

I consider 'nature study' as a study of plant and animal life at first hand, rather than from books; seeing, examining and studying a plant or animal, how it grows; if an animal, how it moves, runs, walks, flies, swims, how it gets its livelihood; and then the child can learn to observe its relation to the life about it and to the world around. Let him observe, for example, ants, the difference between the males, females and workers, how the workers live and care for the colony. He may see a train of ants; let him follow the train off to the nest. Then there are the nests and working habits of wasps and bees.

A student of 'nature study'—a boy or girl—should raise caterpillars to the chrysalis and moth or butterfly state. Collecting, feeding them, watching them through their transformations, is a first class lesson for a child in nature study. So a boy or girl can get a first lesson in physical geography and geology by studying a sand heap or clay bank after a rain—or the work done by a stream or brook.

Nature study is the first step towards natural science, and is all-important in leading one to observe, experiment and reason from the facts he sees. It is of prime importance in teaching a child *what a fact is* in these days of Christian Science and other fads.

A. S. PACKARD.

BROWN UNIVERSITY.

I do not believe I can give in a few sentences my views as to what constitutes nature study. I think the thing is in a chaotic state at present, and I do not feel competent to define it. I have fairly definite ideas as to what material in botany should be included, but botany is only one of the phases of the subject as handled. I think the name nature study is too indefinite to be retained.

JOHN M. COULTER.

UNIVERSITY OF CHICAGO.

I have your letter asking for my definition of 'nature study.' I hope you will succeed in getting this much-abused term properly defined.

I would have nature-study mean the study of living things to determine their habits, instincts, adaptations and relations to environments. To be nature study in the highest sense of the term, the work must be carried on under natural, as opposed to artificial, conditions.

If a broader interpretation were given, where can we stop short of geology, mineralogy, chemistry, physics, and in fact nearly everything else outside of mathematics.

C. P. GILLETTE.

FORT COLLINS, COLO.

Much that has been taught under the name of nature study is not properly a study of nature, but a *memoriter* drill or an empirical abstract of what some one else has learned by a study of nature. The subject has too often been presented under the guidance of teachers who themselves have made no real study of nature—who have no clear understanding of the scientific method of study by which alone matters of natural fact can be approached, and who have not sufficient competence to carry on the study of nature by themselves. But nature study is sometimes what it ought to be: a truly scientific and well-conducted study of nature, of a grade, whether elementary or advanced, appropriate to the age of the pupils; as logical as geometry and as disciplinary as Latin, but entirely unlike either of these standard subjects.

Direct observational appeal to natural phenomena should always be the essential foundation of a real knowledge of nature, and much skill should be exercised by the teacher in selecting from nature's inexhaustible store such phenomena for study as shall really be within reach of the pupils' own observation and understanding. The text-books should serve chiefly to broaden the knowledge gained through observation by presenting additional examples of similar phenomena from various parts of the world. At the same time, and always in a measure appropriate to the grade of the class, the various other processes of scientific method should be brought into play: generalization, invention of explanations, test of explanations by deduction, appeal to experiment, the need of a critical and unprejudiced judgement in reaching conclusions, revision of work and suspension of judgement in doubtful cases. Elementary examples of all these processes may be presented, though those just named are more appropriate than the others for young classes.

In the illustration of nature study with excerpts from poems, I have comparatively little interest, especially when, as is so often the case, the excerpts are not chosen by the teacher, and still less when the teacher's temperament is not poetic. Spontaneous quotations from any field of really good literature in prose or poetry, brought in because of real literary feeling on the teacher's part, are in just measure admirable aids to study of all kinds; but if poems on nature be made an essential part of nature study, it is likely to become emotional rather than scientific and disciplinary.

Desire and capacity to carry the study of nature further should be the chief end of nature study, and it is for this reason that I would emphasize in all grades the disciplinary rather than the sentimental view of the subject. The scientific method should be constantly inculcated, but more by example than by precept.

This should lead to a clear understanding of the order of nature, based not on authority but on the cultivation and use of a keen, unprejudiced, sympathetic reason: emotional sentiment, a subject responsive in so far as it is excited by natural phenomena, is better cultivated in the appreciative study of art and literature than in nature study.

W. M. DAVIS.

CAMBRIDGE, MASS.

Properly it is simply synonymous with the good old term 'natural history.'

As I take it, all zoologists, botanists, biologists, etc., are pursuing 'nature study,' each in his own way. I have no sympathy with the desire of some superficial persons to limit such a term to kindergarten work in zoology and botany, which is about the idea held in some schools.

That kind of work is right and proper and useful in its place, but why should it monopolize the term 'nature study' is known only to the minds of those who can go no farther than the a b c of science.

E. A. VERRILL.

NEW HAVEN, CONN.

I should say that, on the positive side, any direct contact with natural objects, continued by critical or comparative studies, either elementary or advanced, should come under the head of nature study. Negatively, I should exclude all fairy stories about animals and plants, all fantastic stories of creatures more or less imaginary, and should restrict the term so as to include only such work as would bring the student face to face with realities. The essential virtue of nature study lies in its reality, as distinguished from the conventional, artificial or second-hand kinds of learning.

DAVID STARR JORDAN.

STANFORD UNIVERSITY, CALIF.

I should say that by nature study a good teacher means such study of the natural world as leads to sympathy with it. The keynote, in my opinion, for all nature study is sympathy. Such study in the schools is not botany; it is not zoology; although, of course, not contravening either. But by nature study we mean such a presentation, to young people, of the outside world that our children learn to love all nature's forms and cease to abuse them. The study of natural science leads, to be sure, to these results, but its methods are long and have a different primary object.

THOMAS H. MACBRIDE.

UNIVERSITY OF IOWA.

Besides the letters above, a brief quotation is here given from an excellent book recently published by Clifton F. Hodge, Ph.D., of Clark University:

Nature study is learning those things in nature that are best worth knowing, to the end of doing those things that make life most worth the living.

My point is that nature study, or elementary science, for the public school ought to be all for *sure human good.*

Here is a paragraph from a recent letter from Mrs. J. M. Arms, who is in charge of nature study in the schools of Boston, Mass.:

Nature study is simply the study of nature, not the study of books. It is a course of nature lessons especially adapted for elementary schools. Minerals, rocks, plants and animals are the necessary materials for such lessons. The method of study may be expressed in three words, observation, comparison, inference. The child must be made to see the object he looks at, and to this end he tries to draw it and to describe it in writing. Comparative work is mental training, which, combined with the observational training already spoken of, gives a certain degree of mental power. This power gained in the early years increases with continued effort. Fortunately, this work is recognized as one of the potent agencies in producing efficient men and women equipped for a life work that shall make for the betterment and enlightenment of humanity.

[From Cornell Nature-Study Leaflets, *Albany: J. B. Lyon Company, 1904, pp. 11–15.]*

WHAT IS NATURE-STUDY?*

[BY LIBERTY HYDE BAILEY]

NATURE-STUDY, as a process, is seeing the things that one looks at, and the drawing of proper conclusions from what one sees. Its purpose is to educate the child in terms of his environment, to the end that his life may be fuller and richer. Nature-study is not the study of a science, as of botany, entomology, geology, and the like. That is, it takes the things at hand and endeavors to understand them, without reference primarily to the systematic order or relationships of the objects. It is informal, as are the objects which one sees. It is entirely divorced from mere definitions, or from formal explanations in books. It is therefore supremely natural. It

---

* Paragraphs adapted from Teachers' Leaflet, No. 6, May 1, 1897, and from subsequent publications.

trains the eye and the mind to see and to comprehend the common things of life; and the result is not directly the acquiring of science but the establishing of a living sympathy with everything that is.

The proper objects of nature-study are the things that one oftenest meets. Stones, flowers, twigs, birds, insects, are good and common subjects. The child, or even the high school pupil, is first interested in things that do not need to be analyzed or changed into unusual forms or problems. Therefore, problems of chemistry and of physics are for the most part unsuited to early lessons in nature-study. Moving things, as birds, insects and mammals, interest children most and therefore seem to be the proper objects for nature-study; but it is often difficult to secure such specimens when wanted, especially in liberal quantity, and still more difficult to see the objects in perfectly natural conditions. Plants are more easily had, and are therefore usually more practicable for the purpose, although animals and minerals should be no means be excluded.

If the objects to be studied are informal, the methods of teaching should be the same. If nature-study were made a stated part of a rigid curriculum, its purpose might be defeated. One difficulty with our present school methods is the necessary formality of the courses and the hours. Tasks are set, and tasks are always hard. The best way to teach nature-study is, with no hard and fast course laid out, to bring in some object that may be at hand and to set the pupils to looking at it. The pupils do the work,—they see the thing and explain its structure and its meaning. The exercise should not be long, not to exceed fifteen minutes perhaps, and, above all things, the pupil should never look upon it as a "recitation," nor as a means of preparing for "examination." It may come as a rest exercise, whenever the pupils become listless. Ten minutes a day, for one term, of a short, sharp, and spicy observation lesson on plants, for example, is worth more than a whole text-book of botany.

The teacher should studiously avoid definitions, and the setting of patterns. The old idea of the model flower is a pernicious one, because it does not exist in nature. The model flower, the complete leaf, and the like, are inferences, and pupils should always begin with things and phenomena, and not with abstract ideas. In other words, the ideas should be suggested by the things, and not the things by the ideas. "Here is a drawing of a model flower," the old method says; "go and find the nearest approach to it." "Go and find me a flower," is the true method, "and let us see what it is."

Every child, and every grown person too, for that matter, is interested in nature-study, for it is the natural way of acquiring knowledge. The only difficulty lies in the teaching, for very few teachers have had experience in this informal method of drawing out the observing and reasoning powers of the pupil without the use of text-books. The teacher must first of all feel in natural objects the living interest which it is desired the pupils shall acquire. If the enthusiasm is not catching, better let such teaching alone.

Primarily, nature-study, as the writer conceives it, is not knowledge. He would avoid the leaflet that gives nothing but information. Nature-study is not "method." Of necessity each teacher will develop a method; but this method is the need of the teacher, not of the subject.

Nature-study is not to be taught for the purpose of making the youth a specialist or a scientist. Now and then a pupil will desire to pursue a science for the sake of a science, and he should be encouraged. But every pupil may be taught to be interested in plants and birds and insects and running brooks, and thereby his life will be the stronger. The crop of scientists will take care of itself.

It is said that nature-study teaching is not thorough and therefore is undesirable. Much that is good in teaching has been sacrificed for what we call "thoroughness,"—which in many cases means only a perfunctory drill in mere facts. One cannot teach a pupil to be really interested in any natural object or phenomenon until the pupil sees accurately and reasons correctly. Accuracy is a prime requisite in any good nature-study teaching, for accuracy is truth and it develops power. It is better that a pupil see twenty things accurately, and see them himself, than that he be confined to one thing so long that he detests it. Different subjects demand different methods of teaching. The method of mathematics cannot be applied to dandelions and polliwogs.

The first essential in nature-study is actually to see the thing or the phenomenon. It is positive, direct, discriminating, accurate observation. The second essential is to understand why the thing is so, or what it means. The third essential is the desire to know more, and this comes of itself and thereby is unlike much other effort of the schoolroom. The final result should be the development of a keen personal interest in every natural object and phenomenon.

Real nature-study cannot pass away. We are children of nature, and we have never appreciated the fact so much as we do now. But the more closely we come into touch with nature, the less do we proclaim the fact abroad. We may hear less about it, but that will be because we are living nearer to it and have ceased to feel the necessity of advertising it.

Much that is called nature-study is only diluted and sugar-coated science. This will pass. Some of it is mere sentimentalism. This also will pass. With the changes, the term nature-study will fall into disuse; but the name matters little so long as we hold to the essence.

All new things must be unduly emphasized, else they cannot gain a foothold in competition with things that are established. For a day, some new movement is announced in the daily papers, and then, because we do not see the head lines, we think that the movement is dead; but usually when things are heralded they have only just appeared. So long as the sun shines and the field are green, we shall need to go to nature for our inspiration and our respite; and the need is greater with every increasing complexity of our lives.

All this means that the teacher will need helps. He will need to inform himself before he attempts to inform the pupil. It is not necessary that he become a scientist in order to do this. He goes as far as he knows, and then he says to the pupil that he cannot answer the questions that he cannot. This at once raises him in the estimation of the pupil, for the pupil is convinced of his truthfulness, and is made to feel—but how seldom is the sensation!—that knowledge is not the peculiar property of the teacher but is the right of anyone who seeks it. Nature-study sets the pupil to investigating for himself. The teacher never needs to apologize for nature. He is teaching merely because he is an older and more experienced pupil than his pupil is. This is the spirit of the teacher in the universities to-day. The best teacher is the one whose pupils the farthest outrun him.

In order to help the teacher in the rural schools of New York, we have conceived of a series of leaflets explaining how the common objects can be made interesting to children. Whilst these are intended for the teacher, there is no harm in giving them to the pupil; but the leaflets should never be used as texts from which to make recitations. Now and then, take the children for a ramble in the woods or fields, or go to the brook or lake. Call their attention to the interesting things that you meet—whether you yourself understand them or not—in order to teach them to see and to find some point of sympathy; for every one of them will some day need the solace and the rest which this nature-love can give them. It is not the mere information that is valuable; that may be had by asking someone wiser than they, but the inquiring and sympathetic spirit is one's own.

The pupils will find their regular lessons easier to acquire for this respite of ten minutes with a leaf or an insect, and the school-going will come to be less perfunctory. If you must teach drawing, set the picture in a leaflet before the pupils for study, and then substitute the object. If you must teach composition, let the pupils write on what they have seen. After a time, give ten minutes now and then to asking the children what they saw on their way to school.

Now, why is the College of Agriculture at Cornell University interesting itself in this work? It is trying to help the farmer, and it begins with the most teachable point—the child. The district school cannot teach technical professional agriculture any more than it can teach law or engineering or any other profession or trade, but it can interest the child in nature and in rural problems, and thereby join his sympathies to the country at the same time that his mind is trained to efficient thinking. The child will teach the parent. The coming generation will see the result. In the interest of humanity and country, we ask for help.

How to make the rural school more efficient is one of the most difficult problems before our educators, but the problem is larger than mere courses of study. Social and economic questions are at the bottom of the difficulty, and these questions may be beyond the reach of the educator. A correspondent wrote us the other day that an old teacher in a rural school, who was receiving $20 a month, was underbid 50 cents by one of no experience, and the younger teacher was engaged for $19.50, thus saving the district for the three months' term the sum of $1.50. This is an extreme case, but it illustrates one of the rural school problems.

One of the difficulties with the rural district school is the fact that the teachers tend to move to the villages and cities, where there is opportunity to associate with other teachers, where there are libraries, and where the wages are sometimes better. This movement is likely to leave the district school in the hands of younger teachers, and changes are very frequent. To all this there are many exceptions. Many teachers appreciate the advantages of living in the country. There they find compensations for the lack of association. They may reside at home. Some of the best work in our nature-study movement has come from the rural schools. We shall make a special effort to reach the country schools. Yet it is a fact that new movements usually take root in the city schools and gradually spread to the smaller places. This is not the fault of the country teacher; it comes largely from the fact that his time is occupied by so many various duties and that the rural schools do not have the advantage of the personal supervision which the city schools have. [. . .][5]

[From The American Monthly Review of Reviews, *vol. 23, no. 4, April 1901, pp. 463–464.]*

NATURE-STUDY ON THE CORNELL PLAN.

By Professor L. H. Bailey.
(Of Cornell University.)

A PREVAILING tendency in education is towards nature and naturalness. That part of the movement which looks to things afield for its inspiration is usually known as nature-study. This term may mean anything or nothing. There is no uniform body of principles or practice included in the term. The greater part of what is called nature-study is merely easy or diluted science. Another part of it is sentimental affectation. Between the two should lie the real and true nature-study—that which opens the eyes of the child to see nature as it is, without thought of making the child a scientist, and without the desire to teach science for the sake of science. The nature-study of the scientist is often the mere interpretation of scientific fact and discovery; but the child receives this knowledge second-hand, and what it receives is foreign to its own experiences. The gist of such teaching is to impart knowledge, but the true nature-teaching seeks rather to inspire and to enlarge one's sympathies; mere facts are secondary. Every person lives always in an environment: if he do not have a spontaneous interest in that environment, his life is empty. We live in the midst of common things.

The Cornell nature-study movement seeks to improve the agricultural condition. It wants to interest the coming man in his natural environment, and thereby to make him content to be a countryman. This is the only fundamental solution of the so-called agricultural question. All things hinge on the intellectual effort and the point of view of the individual.

The first effort was to teach the teacher in the rural district school; but this teacher is hard to reach. She is removed from associations and conventions. She is the teacher of least experience, and frequently of least ambition. She follows. It soon became apparent that the leaders must first be reached. In the largest cities of New York State, the agitation bore its first fruits. The country places are now taking it up. Before the movement was definitely organized, many rural schools were visited. The teachers were found to be willing to introduce a little sprightliness and spontaneity into their work, but they did not know how. They wanted subject-matter. The children were delighted with the prospect of learning something that had relation to their lives.

Readable leaflets were prepared on living, teachable subjects, for the purpose of giving the teacher this subject-matter and the point of view. It was not desired to outline methods, for methods are not alive. If the teacher were awakened and were given the facts, the teaching would teach itself. The first constituency was secured by sending an instructor or lecturer with the State teachers' institutes,—for the State Department of Public Instruction kindly made this possible. From teacher to teacher the idea spread. Now 17 leaflets have been issued and about 26,000 teachers are on the mailing-list by their own request.

The leaflet attempts nothing more than to say something concise and true about some common thing, and to say it in a way that will interest the reader. The point of view is the reader rather than the subject-matter. The leaflets aim to send the reader to nature, not to record scientific facts. The first leaflet was entitled "How a Squash Plant Gets Out of the Seed." A botanist said that the title was misleading: it should have read, "How the Squash Plant Gets Out of Its Integument." Herein is the very core of the whole movement: it stands for "seed," not for "integument."

How is the teacher to use these publications? As he will. It is recommended that he catch their spirit, and then set the pupils to work on similar problems. It is not designed that the matter be made a part of the curriculum, for then there is danger that it may become perfunctory. Nature-study should supply the enthusiasm of the schoolroom. Nor is it enough that the leaflets are published and sent to applicants. They are followed up by personal correspondence and advice. A leaflet is never out of date if it is worth printing. It is used over and over again, year after year, and becomes more useful the longer it is used.

But there must be something more than mere intellectual assent to induce the teacher to take up the nature-work. The teacher is tired and brain-weary; but ten or fifteen minutes a day given to plant or bird, or bug or brook, enlivens the whole school and makes the eyes sparkle. More than this, the subject becomes the theme for the English compositions, and one of the bugbears of the schoolroom vanishes. Writing is easy when the child writes naturally of what it knows.

The second distinct movement in this nature-study enterprise was the organization of the children into what are called Junior Naturalist clubs. Already there are 1,100 clubs, with a total enrolled membership of over 30,000 children. The idea is to get the children to do something for themselves. The club is theirs. The teacher is asked if she will encourage the organization of one or more clubs in her school. She suggests it to the children and leaves it with them. They meet and organize, and send the names of the members and officers to the Nature-Study Bureau, at Ithaca. The club is named by its members. It may be "The Bright Eyes," "The Wide-Awakes," "The Investigators," or named for the village or the teacher.

Each member pays dues twice each month; this payment consists of an essay or letter on what has been learned of nature-life. This payment may be made by the very essay which the pupil wrote in its composition period. To the home office they come by the hundreds, and the children are encouraged to write as they think and feel. "Corrected" essays are not desired. Each payment of dues is checked up on the member's personal card, and those who meet their obligations promptly receive a neat "Junior Naturalist" button.

The children are guided in what they are to see. There is published a "Junior Naturalist Monthly," which suggests the work for the month. So far as practicable, these monthlies take up the topics that have been expounded in more detail in the teacher's leaflets; for the teacher thereby is brought into more intimate touch with the work of the children. The monthly lesson may be on seed-travelers, birds and bird-houses, an insect, a plant, a toad, a spring brook, or other practicable and vital topic.

In this "Junior Naturalist" work, the teacher has only a supervisory interest. She is not asked to take up new duties and responsibilities. The children manage their own affairs. A most gratifying result of the Junior Naturalist enterprise is the aid that it renders in school discipline. Naturally, the members have pride in their club and its standing. The club has meetings, as a rule, and discusses the lessons. It is conducted on parliamentary principles. Teachers are beginning to testify to the disciplinary value of the children's clubs, and to suggest that instructions in "rules of order" be made a part of the work. By appealing to the club spirit, the teacher is able to improve the *morale* of the school without conscious effort on her part; and the main purpose of the movement—to quicken the pupil's interest in the things with which he lives—is forwarded at the same time.

The immediate correspondence with the Junior Naturalists is in the hands of a judicious and sympathetic man of affairs, who is known to the 30,000 children as "Uncle John." To him they may write with confidence and freedom; and to receive a letter from him is regarded as an experience. A useful feature of the work is the encouragement of correspondence between widely separated clubs. The letters or dues of a city club may be exchanged with those of a country club. Some of the dues take the form of drawing-work, which may have been a part of the regular drawing period of the schoolroom. These drawings are useful for exchange. The drawings of leaves and of "Jack Frost" have been among the most useful. If the monthly lesson is on "Apple Twigs," or any other topic that is somewhat foreign to the city child's life, the country clubs are asked to collect specimens and to send them to their city correspondents. This is an obligation that is joyfully rendered. Although this nature-study movement is a New York State enterprise, outside clubs have not been refused. Some of these clubs are in foreign countries. There is one in Egypt, and another in Tasmania. They are scattered over the Union. This wide range adds greatly to the value and interest of correspondence and interchange, although it will be necessary to curtail the outside work in the future.

[From The Century Magazine, *vol. 74, no. 6, October 1907, pp. 960–967.]*

THE COMMON SCHOOLS AND
THE FARM-YOUTH

BY L. H. BAILEY
Director of the College of Agriculture,
Cornell University

IN three previous papers* I have discussed three phases of the outlook on agriculture as expressed by students—why certain young persons desire to leave the farm, why others desire to remain or even to remove there from town, and what the agricultural college is doing for the farm-youth. It now remains to complete the series by a discussion of what the common school can do for the farm-youth.

The agricultural colleges are now accomplishing results of great and permanent value, in spite of the fact that they are isolated from the common schools, on which good collegiate training is supposed to rest. The agricultural country is well peopled with good farmers, in spite of the fact that the common school in the open country has given them no direct aid in their business.

---

* See THE CENTURY for July, August, and September, 1906.

Sympathy with any kind of effort or occupation, and good preparation for engaging in it, are matters of slow and long-continued growth. This growth should begin in childhood, and should be aided by the home and the school. The country school carries a greater responsibility than the city school, in proportion to its advantages, for it is charged not only with its own country problems, but with the training of many persons who swell the population of cities. The country school is within the sphere of a very definite series of life occupations.

We may well begin our discussion of some phases of the rural-school problem by stating two propositions: (1) education should develop out of experience; (2) the school should be the natural expression of its community.

The country schools—I now make no reference to other schools—do not exhibit either of these principles. The subjects taught in them are not the essentials; the school does not represent or express the community. I do not know that any schools teach the essentials, except as incidents or additions here and there, and essentials cannot be taught incidentally or accidentally. Arithmetic and like studies are not essentials, but means of getting at or expressing the essentials. The first effort of the school should be to teach persons how to live.

The present methods and subjects in the rural schools have come to the schools from the outside. If we begin the school work with the child's own world, not with a foreign world or with the child's world as conceived of or remembered by the teacher or the text-book maker, it is plain that we have by that very effort started a revolution.

In making these remarks, I do not lose sight of the fact that we are making distinct progress in these very directions, or that teachers recognize the need of a change in point of view. Perhaps the best way to discuss the subject is to comment on what is already beginning to be accomplished. This will show the direction in which we are trending.

The Status of the Nature-Study or Experience-Teaching Movement

EXPERIENCE-TEACHING has now come to be one of the conspicuous phases of current educational work, and it is an interest that also attaches strongly to the rising feeling for release from conventionalism. It is expressed in kindergartens, manual-training, and the like, and in that teaching of natural history to which the term nature-study is commonly restricted. Most discussions of nature-study consider only its technical phases as a school exercise, dealing largely with the subject-matter, and the methods of teaching. In such discussions it is difficult for the layman to catch the spirit of it. In reality, the nature-study interest is one of the expressions of an underlying and redirecting tendency in our development. It is unfortunate not to know its philosophy, for we miss its significance. The movement is gradually being accepted as a necessary and abiding direction in education; here and there a teacher has worked the philosophy into practice, and schools have found a place for some expression of it in the scheme of studies. The growing sentiment and experience are now being reflected in syllabi and courses of study. More than forty States, Territories, and Provinces have officially recognized nature-study or its closely associated subjects. Sometimes this recognition is the publication of a State course of study, sometimes the adoption of a text-book or recommendation of literature, sometimes the dissemination of leaflets, or, again, the passing of a mandatory law by the legislature. State policies are necessarily conservative, so that this wide recognition means that nature-study is at last fully established in the public confidence.

Perhaps the best treatise on nature-study that could now be made for teachers would be a skilful editorial combination and discussion of the various printed courses of study. My present purpose, however, is to try to determine what are the current conditions and tendencies as expressed by these syllabi. The discussion divides itself into two parts: (1) Objects of nature-study; (2) Methods of nature-study.

Some Objects of Nature-Study

A STUDY of the various expressions in the State and other syllabi as to subjects of nature-study suggests a number of fairly definite summaries. Some of these conclusions may be stated as follows:

It is the purpose of nature-study to develop the child's native interest in himself and his surroundings. It proceeds on the theory that the best educational procedure with the young is first to direct the personal sentiments, powers, and adaptabilities. Of course we must consider not only what the child's interests and powers are, but also how we can aid him to grow into a man; but we cannot annihilate the native adaptabilities without endangering the child. It may be even dangerous to try to suppress them. If these tendencies and sentiments are not directed, they are likely to develop into wild and wasteful energies. The causes of truancy lie in part in our over-diligent efforts to repress the native enthusiasm of the child. A good part of our training of children, I fear, is expressed merely by the command "Don't." Each truant is a problem in himself, but it is probable that most truants belong to one or another of three classes: (1) the vicious class; (2) the low mentality class; (3) the class that will not conform to usages and to customs, and in which the energies tend to run riot, or at least to express themselves in erratic and unconventional ways. These last are the true truants. They are repressed children. A child of this class may be likened to a jack-in-the-box: he is forced into conventional limits, but is always ready to break out in a way that brings consternation to the well-behaved.

Nature-study, therefore, is to begin with general, common, normal, and undissected objects and phenomena, rather than with definition and classification, in order that the child may be developed natively. Definition and classification are the results of the accumulation of experience. They are not primary educational means or methods. Definition always lags behind knowledge. It is likely to take the place of knowledge in the child's mind. It did in the old botany and grammar and physiology. As soon as we begin to compress knowledge and experience into the limits of definition, we take away the life, spontaneity, and enthusiasm of it. Definitions are for mental guidance after experience has accumulated, and they become more exact with the maturity of the person. No doubt we have over-defined the subject-matter in our text-books.

Nature-study is coming more and more to be an out-of-door subject, for the child's interest should center more in the natural and indigenous than in the formal and traditional. It is not our sphere to live chiefly in buildings. Nature-study began very largely with object-lesson work. The objects might have been collected out of doors, but they were taken into the school-room to be studied. This was a distinct advance over the older type of object-study, because it tended to substitute natural objects for artificial and geometrical and unpersonal ones; but it did not develop into true nature-study until a distinct effort was made to study the objects and the phenomena just where they occur in their normal relationships. There can be no effective sustained nature-study when the work is confined in a building.

One is impressed in the various expressions coming from many parts of the country with the universality and unanimity of the nature-study movement, indicating the existence of a general feeling that the schools are not adequate and not vital.

The nature-study teaching has introduced many new and significant phrases into the teacher's vocabulary, as, for example, "increasing the joy of living," "sympathetic attitude toward nature," "increased interest in the common things," "to train the creative faculties."

The keynote of nature-study is to develop sympathy with one's environment and an understanding of it. The long-continued habit of looking at the natural world with the eyes of self-interest—to determine whether plants and animals are "beneficial" or "injurious" to man—has developed a selfish attitude toward nature, and one that is untrue and unreal. The average man to-day contemplates nature only as it relates to his own gain or personal enjoyment.

The end of nature-study is to develop spiritual sensitiveness and insight; therefore, it must not cease with mere objects and phenomena. In this it differs from the prevailing conception of science-teaching. I think that I catch this note in the syllabi and books that I have examined. This attitude accepts phenomena as real, and regards what we call "progress" to be really such. It accepts the world as good. It does not depreciate the need and importance of introspection, but regards introspection and meditation as exercises for a mature and maturing mind, and holds that such exercise is most effective when most closely related to experience. Nature-study is not merely objective if it is developed in the way in which it should be developed. If we develop first the meditative, passive, and subjective habit, then we are oriental; but the spirit of the West is to live actively with the world.[6]

Methods of Nature-Study

A STUDY of the various statements in the syllabi of methods of nature-study teaching warrants a number of significant conclusions, some of which are as follows:

The methods are coming to be somewhat concrete because the motive is being understood. The motive of nature-study work is reaction from formalism and from method; it is revulsion against the introduction of technical laboratory methods at a too early age; it is revolt from the spirit of grown-up scientists as applied to elementary educational work.

What method there is in the work is characterized by spirit. It embodies the spirit of individuality, spontaneity, enthusiasm. It is essentially informal and undogmatic. It arouses human interests. The old educational procedure seemed to be to try to make children as like as two peas. In fact, this procedure is still in vogue, and this accounts for much of the deadness of school work. Individuality and personality, however, are the primary considerations in education, and the nature-study method aims to develop them. It puts a premium on original modes of apprehending knowledge. It develops personal responsibility and initiative.

In practice, nature-study develops many new modes of expression, as action, writing, speaking, drawing, color, music. That is, it develops the whole person. It also leads to a fine feeling for poetic interpretation.

These ways of procedure tend to make the school a unit instead of a mere assemblage of classes. They break up the monotony and formality of the curriculum, and tend to give the school an expression of naturalness. They add variety and vividness. Nature-study should correlate and inoculate all school work. It puts a new motive and meaning into the school by making the school real and giving the teaching local application.

Nature-study practice broadens the meaning of schooling. Consider the scope and breadth of the subjects that it touches: plants, animals, weather, the sky, fields and soils, health, affairs. The fact that so many subjects are touched is one reason why teachers of science are likely to disparage nature-study, for these teachers pursue one subject continuously for a considerable time and in much detail; but the science-teaching of the college and the best high schools must not dictate the subjects and methods of the elementary schools. No one teacher is likely to cover all the subjects coming within the denomination of nature-study. The fact that so many subjects fall within its sphere allows of choice, and thereby adds all the more to the spontaneity and significance of school.

The nature-study method marks the final rise of the school-garden as a central means in elementary school work. This is likely to be the pivot about which personal nature-work revolves, because it is near at hand, concrete, and controllable. It is the laboratory from which all enterprises diverge. In time it will come to be regarded as one of the essential parts of any elementary school. It provides relevant laboratory work; and no school, from the kindergarten to the university, is a good school unless it has laboratory work. Kindergarten-work, manual-training, nature-study, are all laboratory work.

The methods of nature-study tend to connect the school with the home. They make schooling a serious affair. The school becomes a social force. Every teacher has felt that a good part of the patrons of any school look on the teacher with a sort of self-complacent and patronizing air, as if they did not take the teacher's work to be serious or really important. They seem to accept it as one of the things that custom has imposed. I heard a school-gardener say that the parents in her district were silently opposed to the school. When asked for the reason, one of them remarked, "If the children go to school, we can't make them do anything for us." That is, there are two opposed forces, the school and the parents. The school-garden and other nature-study work tend to correct this antagonism or separateness. In many cases the individual school-garden may be in the home grounds rather than on the school grounds. The school, the home, the community are only different phases or expressions of experience. The redirected school will develop the economic and social consciousness.

We may sum up this review of methods by saying that the teaching begins with the actual, the tangible, the significant. We do not begin with classifications or systems, or with the idea of giving the child a complete view of a subject. We deal with the concrete. The pupil will gather experience and gain wisdom, and finally, we hope, come to systematic knowledge. We shall not teach merely for the purpose of giving information: that can be got in a book. In the elementary grades in a country school, I think we shall do far better to teach the raising of a crop of corn, or the making of butter, than the principles of tillage, of soil fertility, or the theory of feeding cows. We should begin to teach by specific cases and examples. Possibly in the high schools we can begin to teach principles of soil fertility and cattle feeding, but there is danger of going too far in these abstractions even there.

Some day the common schools will prepare for colleges of mechanic arts and agriculture as consciously as they now prepare for literary colleges. It is a question whether the proper demarcation between the common school work and the college work will not then lie in the school dealing with actual problems and the college dealing also with the theories

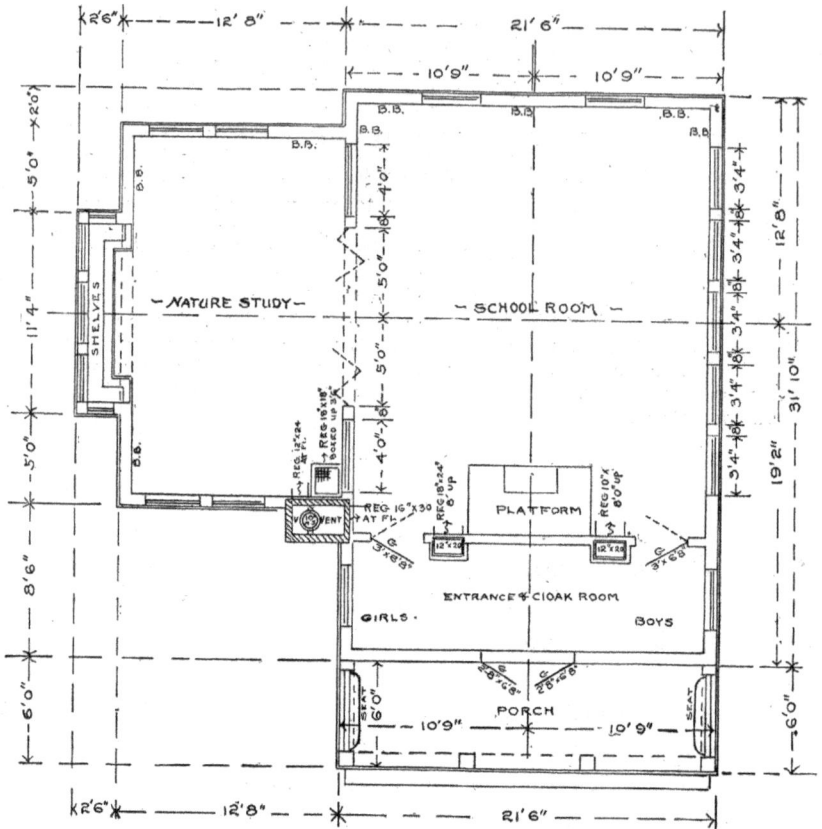

Figure 25. Floor plan of the Cornell Rural School-House, drawn by W. C. Baker.

and the classified science. I think that the syllabi for agriculture in high schools err in covering the same ground that college courses cover, only in a more elementary way. Probably they would do better to confine themselves more closely to special problems that mean something to the pupil and the community. It is not at all necessary that the high-school pupil should "develop a subject" or have "a body of knowledge." Much of the present high-school work is far beyond the pupil. The people have always asked for concrete knowledge and training.

Figure 26. Front view of the Cornell Rural School-House, drawn by Philip B. Whelpley, from a photograph. The College of Agriculture at Cornell erected this small rural school-house on its grounds, to serve as a model, and to house a real rural school as part of its nature-study department. School-gardens and play grounds have been made at one side. The building, furniture and supplies cost $1,983.31.

It is a question whether nature-study and crafts subjects should be taught in the grades merely to illustrate or vivify some literary text or story. For example, is it worth while to exemplify Robinson Crusoe by studies of dogs and parrots and the making of canoes, as is now the vogue? These exercises are really extraneous, after all, and a kind of acting. It is a question whether it is profitable for a mere child in the grades to build canoes unless the exercise comes naturally as a part of personal experience. The object-work to illustrate literature lessens and subordinates the meaning of the object; and I cannot help feeling that the effort might much better be expended on objects for their own sakes, and that have relation to experience, letting literature be taught in some other way. We do not need any excuse for the study of nature.

Results to Be Expected from Nature-Study Teaching

Persons are always asking for the results of the nature-study work, as if they expect that statistics can be given in reply. They want to know how many teachers are teaching it, how many children are interested in it, how many school-gardens there are, how many syllabi are in use, how many pupils are enrolled, and the like. All this is well in its way, and is important, though the results of nature-study are not to be measured by these formal means, but, rather, by a general elevation in the mode and tone of the school, and in the point of view of the community. The school must be reorganized to meet the child's needs. It must be simplified. Subjects must be taken out, rather than put in; but whatever subjects remain, the nature-study philosophy and point of view must run through them all, for it is a fundamental educational means. Most of the criticisms of nature-study are made against what are thought to be faulty methods here and there. It may be a question whether these criticised methods really are faulty; but even if they are, and if all the work has been inadequate, nevertheless the nature-study movement will abide. It is one expression of the new education.

If this experience-teaching is so fundamental, we must not look for results quickly. Spiritual movements proceed slowly. It may require a generation yet to get us out of the habit of teaching merely the names of things.

It has been said that the current movement toward nature-study is misdirected, since all human activities, of whatever kind, proceed from experience. Language, for example, is only a means of expressing experience; therefore Greek study is nature-study. However, the evolution of a language is the experience of a race; what we now argue for is the using of the experience of the individual. Of course no one would advise against the use of race-experience, as expressed in language and literature; but education should begin with the person, which is the concrete.

It is a fallacy to consider that nature-study must be merely correlated with the present school subjects except as a means of starting and establishing its spirit. Nature-study teaching is a way of conducting the school work so that it will have personal application and meaning.

The school must be given a new purpose or expression. Our school systems are now really developed for the few—for those who are good

Figure 27. Rear, and work-room end of the Cornell Rural School-House, drawn by Philip B. Whelpley, from a photograph.

"scholars." Other pupils are expected to emulate these few, whereas they may have a wholly different order of ability. When education becomes personal, all this will change. Well-developed experience with one's normal environment is nature-study: it lies deeper than the adding of a subject to the course, deeper than merely to be "correlated with." It is quite the opposite of "correlation with," as if it were applied from the outside: it is giving direction to, making application of.

Application to the Country School

Just now nature-study is the stepping-stone to the introduction of agricultural studies. This is an indication that it is a means of connecting the school with the real life and activity of the community; but nature-study is a means of preparing the pupil for all kinds of school work and for all places, as well as for agriculture and for the country. It is a redirecting agency. In time, as the schools develop, we shall find that we shall not need to introduce agriculture as a separate study, even in rural districts, at least not below the high school, for in such districts the whole school effort will have an agricultural, country-life, or nature-study trend.

Lest I be misunderstood, I will say at once that I am not opposed to the introduction of agriculture as a separate study into the elementary rural school. In fact, such introduction may be the very best means of bringing about the deeper and more fundamental re-directing of the school that is essential to its full effectiveness. I look on the separate teaching of agriculture as a present means to an end. We should not lose sight of the fact, however, that the schools are actually being redirected much more rapidly than those not engaged in school work may be aware.

In time, the beginning schools will probably not teach any of the present-day subjects under their present names; but this will adjust itself in the natural course of evolution. The greatest need is to reorganize the teaching of the subjects that are already in the country schools. Geography, for example, will deal first with the local country and its affairs. Of course the methods have changed greatly in a generation; but the old geography was largely of the ballooning variety, beginning with the universe and descending through the solar system to the earth.

All this is rapidly changing. If the school is in the open country, it may give attention to fields, birds, soils, brooks, forests, crops, roads, farm animals, hamlets, and homes. Geography can be so taught in the schools as, in ten years, to start a revolution in the agriculture of any commonwealth.

Arithmetic needs redirecting in the same spirit. The beginnings of a new motive in it are now becoming prominent. The principles of number are the same wherever taught, but practice problems may have local application. These problems have heretofore dealt with theoretical, urban, middleman, copartnership subjects, and sometimes have been mere numerical puzzles. It is significant that the arithmetic problems that the country child takes home do not interest the old folks. This is only because the problems mean nothing to them. Many of the problems of the farmer are numerical—soil moisture, fertility questions, feeding rations, spraying, cost of labor and of producing crops, and all manner of accounts. Number can be so taught in the schools as, in ten years, to start a revolution in the agriculture of any commonwealth.

Reading needs similar reorganization. This is everywhere recognized, and distinct progress is being made. It is not desirable to eliminate the customary types of literature of the masters; but something may be added to make the reading vital and applicable. It is not difficult now to find good pieces of English composition that deal with the customary practices and affairs of the open country, and that point the way to better things. Reading and spelling can be so taught as, in ten years, to start a revolution in the agriculture of any commonwealth.

Even manual-training needs new direction as it touches country life. It may not be necessary to eliminate the formal exercises of model work and weaving and the like; but some of the practical problems of the home and farm may be added. How to make a garden, to lay out paths, make fences and labels, are manual-training problems. How to saw a board off straight, to drive a nail, to whittle a peg, to make a tooth for a hand hay-rake, to repair a hoe, to sharpen a saw, to paint a fence, to hang a gate, to adjust a plow-point, to mend a strap, to prune an apple-tree, to harness a horse,—the problems are bewildering from their very number. Manual-training can be so taught in the schools that are equipped for it as, in ten years, to start a revolution in the agriculture of any commonwealth.

All such teaching as this will call for a new purpose in the school-building. The present country school building is a structure in which children sit to study books and recite from them. It should also be a place in which children can work with their hands. Every school building should have a laboratory room, in which there may be a few plants growing in the windows, and perhaps an aquarium and a terrarium. Here the children will bring their flowers and insects and samples of soil, and varieties of corn or beans in their season, and other objects that interest them, and here they may perform their simple work with implements and tools. Even if the teacher cannot teach these subjects, the room itself will teach. The mere bringing of such objects to school would have a tremendous influence on the children; patrons would ask what the room is for; in time a teacher would be found who could handle the subjects pedagogically. Now we see children carrying only books to school; some day they will also carry twigs and potatoes and animals and stones and tools and contrivances and other personal objects.[7]

My plea, therefore, is that the school accept all wholesome conditions in which it is placed, and that it begin with the sphere in which the child lives. The working out of this philosophy is nature-study (I know of no better term); and this philosophy goes deeper than mere manual-training, or than arts and crafts studies, or than bare "self-activity." Nature-study, as I conceive it, is not another subject, not something external or added to. It is a means of education, internal, central, essential, fundamental. In time nature-study and agriculture will be as much a part of the country school as oxygen is a part of the air.

[From The Nature-Study Review, *vol. 12, no. 6, September 1916, pp. 247–249.]*

WHEN THE BIRDS NESTED

L. H. BAILEY

I was fortunate to have been born and sent forth near a brook and several catholes,* in the forests and with a varied wild life. The wolves had just disappeared as I came into knowledge of my surroundings, but bears and lynxes were now and then seen and deer were not uncommon. Nearby

---

* NOTE—The term cathole seems to be little known at present, as it was used in the early days in Michigan. It is not a hole in the cellar door to let the cat in and out; nor is it a nautical term, as in the dictionaries. It was applied to a small bog or swamp, usually less than an acre in extent, as I recall it, and sometimes only four or six rods across. Commonly it was deep in the center, often with considerable muck deposit. These holes were undoubtedly post-glacial, perhaps in large part the depressions left from the melting of remaining masses of ice. About their edges grew willows, sedges, and other lowland growths, but the hard land came close around them. I have heard it said that they were called catholes because of the cattails that grew in them; and others say it is because they became depositories for departed cats and all other offcasts, but this I doubt; yet there were lots of things in those catholes. L. H. B.

was a wonderful rookery of passenger pigeons, and all my early boyhood was animated by the clouds of flying birds in the feeding seasons. The Indians, migrating with the fishing and the game, were a constant wonder. A mile away was Lake Michigan, and although the roar of it became a part of me and I often ran its shores, it was nevertheless always another world, a great place outside of me, mighty and compelling but yet not within my waking ambitions.

I doubt whether any recent boy feels that old charm of the cathole—of that small swamp with a deep hole in the center, in which everything seemed to grow, where strange birds nested, to which all things retreated, where there was water life beyond reach, and whence a small boy expected everything unearthly to come. It was a part of the pioneer life, how much a part we did not then know for we thought the fever-and-ague to come from the miasma of the newly broken ground. It must have been more of a factor with us than the coulee of the farther Out-West, for it was wet and full of breeds year in and year out. I cannot make the young folk understand that certain dry lands were once the scene of catholes, with perhaps a corduroy road across them and with the logs a-swim in spring, with whelms of peepers when the pussy willows were out, frightsome snakes of all imaginary kinds, and cat-bird nests in the margins. To this day the squall of the cat-bird recalls a cathole! Very well! They have gone with the Indian, the passenger pigeon, the many curious traps concealed in the runways, the burning logs, and the unsolvable mystery of the great woods.[8]

My father's farm was a zoological and botanical garden,—not that it was different from any other farm, but because so many things seemed to live and grow there that I thought I could never find the end of them. To make a list of them, to put down where I saw them and what they did,—this seemed the only way to find out how many they were. This was no easy task, seeing that I did not know the names of them, in the early days, and had little way of finding out except to use such names as the settlers or certain antiquated books applied. Often I wonder whether the joy of the field is so keen in these perfected days when everything is explained so carefully and we are so well instructed in what we ought to see.

Three sets of lists I remember to have kept; one was of the daily weather, one of the birds, and later one of the plants. Very simple were these lists, scarcely to be dignified by the name of note-books, but they served to prolong and to multiply the experiences. Any old account book or composition book, with a few unused leaves, was sufficient. These leaves were carefully ruled up and down into columns for the name of the bird (marvellous names I must have given them!), when it began to build its nest, when completed, the first egg laid, the subsequent eggs unto the last, the period of incubation, when the birds flew, and how many. This was indeed a very simple record, but the number of nests under observation would run into the tens and perhaps more and each one was visited every day as regularly as the other "chores" were carried. It became a sort of game or play with me, and it was part of the game to visit the nests when the birds were away and would not be frightened. Back and forth from cultivating corn or driving the team here and there or following other regular farm work, these nests were home-plates and bases (we did not have base-ball then but only long-ball and two-old-cat), and reason enough to go the long way or the short way. Some few of the old trees still stand, and now, with memory running back to those years, I go to them when I visit the old place and look for the nests and the eggs that are not there. The hollow stumps and rails have vanished years and years ago and I cannot look for the pale eggs of the blue-bird. Nor do I find the nests of the cat-bird or the chewink, and even the wren has left the premises. The day by day "tab" on those few birds became a real part of my life, all the more interesting to me, I fancy, because I knew so little about them from books and had so few ways of finding out. My observations must have been very imperfect; but how real were those birds and how I loved to put down the dates!

[From School Science and Mathematics, *vol. 18, no. 2, February 1918, pp. 99–103.]*

THE SCIENCE ELEMENT
IN EDUCATION.*

BY L. H. BAILEY,
Ithaca, N. Y.

The address was divided into two general parts: First, an expression of opinion and point of view on the traditional division of educational topics into the arts and the sciences; second, the contribution of science teaching to the development of civic ideas, particularly to the achievement of democracy. On the old controversy between the humanities and the sciences, now again revived, he spoke as follows:

> We are born to "things" and to "phenomena." We know things, smell them, wear them, handle them, see them. They comprise the goods of life. The phenomena represent the interplay of forces.
>
> We cannot conceive of existence without things and phenomena. Even our conceptions of the state of immortality are imaginations of glorified material things, even to cities not made with hands. The life of the day

---

* Abstract of address before the Central Association of Science and Mathematics Teachers, at Columbus, Ohio, November 30.

is the life of experience with things. There is wood and objects made of wood; rocks and works in rock; land; trees, birds, quadrupeds, streams, hills; slants and levels and inclines; houses big and little; people; the sky; the light and the dark and the gloaming; machinery; food; fabrics wonderfully fashioned of many wonderful materials; ships and the sea under them; timeless shores; great cities and the vast accumulations in them; things that are and things that have been; action and reaction of materials and of forces; actions present and actions past; movement everywhere, quiescence everywhere; numbers and the relations of numbers; quantities; the human mind. The regulated knowledge of things and phenomena is science.

As we depend on things and phenomena, so is the science of them essential; and what is essential is necessarily educational, if we are to live rationally.

We have confused ourselves by explaining to ourselves that we understand. We build up philosophies on subjective processes, and depart from contact with the things and the phenomena about which we philosophize and psychologize.

We are given to the use of phrases and catch-words. We have said far too much of the value of some subjects as "discipline." We make unnecessary and untrue contrasts of "conventional" and "modern" subjects; of "humanities" and "the arts" and "the sciences." I doubt whether the terminology represents essential differences, or means as much as we think it means.

In view of the experience in life, the effort to prove that educational values inhere somewhat exclusively in certain subjects becomes merely weariness. I think that all knowledge is good for the human mind. I have never known any education to hurt anybody, even though it is said to be poor education. Some educational effort is less effective than others because it is less organized, has less constitution, is founded on less knowledge, dominated by less sense, and propelled by a poorer teacher.

You will understand by these allusions that I am not to discuss educational values as such. Now and then we need to come back to science as science, as a knowledge and appreciation of the life we live. Here are the values that cannot be gainsaid.

I know of no line between science and non-science in education. I know of no "humanities" that are not science. I know of no "science" that is not humanities. If chemistry may be a means of effective education, so may history. Even tradition has educational value, for tradition is part of the natural history of the human race.

In this modern world of intense activity I have no fear of what we call tradition in education, although some of our writers seem to have made it a bogey-man. We might profit greatly by more tradition. It is a vast misfortune to separate ourselves so completely from the human past. We have forgotten our grandfathers and soon we shall forget our fathers.

Old subjects may be more worth while in the schoolroom than the new ones, because they carry with them much accumulated human interest; they may also be better organized as educational agencies: I should want them taught as live subjects, however, not as dead ones.

I deprecate the constant iteration of "science" and "humanities." If we were to cease this useless discussion, magnifying the differences, we should soon forget the division, for the division is arbitrary. Any subject is only what human beings make it, and of any two of the recognized subject-courses one is as humanistic and as cultural as the other were it possible for one teacher to teach the two subjects equally well. The greatest deficiency of the older line of subjects is the assumption that it is superior in itself and has more power of mental training: this attitude foreshortens the reach of the teacher and deprives him of the best approach to his pupil.

The greatest deficiency in the science line of subjects is the assumption that it is superior because it may have direct application to the arts of life: this attitude limits the range of the teacher and forestalls the full meaning of the subject. So completely do we advocate and justify science because of its application that we almost forget that the highest quest of mankind is to apprehend the truth. Recently I sat for two days hearing papers on many natural science subjects, mostly without application to current affairs. It was like a translation into a super-world, into a realm of high endeavor for the sake of the endeavor, beyond politics, commercial drives and compromises. No one paused to ask what it was for, what use it had, what anybody expected to gain by it, or what the public would think of it. I should have felt the same satisfaction had I sat for two days with a body of distinguished classicists. To state facts and conclusions because we think they are true and to let the truth be its own reward is reason enough.

This motive to know the truth and to interpret it is just as evident in what we call the humanities as in what we call the sciences. It comprises a reason for education. It establishes the ideals in the young, for one cannot be right with oneself or be rectilinear in relations with one's fellows unless one thinks first of the integrities rather than the expediencies.

This mental attitude and the intellectual cultivation may be derived from Greek or from geometry or from biology, although the color of the result will differ with the subject and particularly with the teacher and the atmosphere of the instruction. The different subjects develop their own mental aptitudes, one, as mathematics, the integrity of the mental process, another, as geology, accurate observation. Herein lies the great value of modern education, in the fact that we may secure the central result and at the same time stimulate the variation that develops the mind sympathetically and that opens it to the vast satisfactions of life. I would not want the pupil or the student to be educated in Greek alone or in geometry alone or in biology alone.

I do not like the course of study that is all of the kind that we loosely call "the classics"; no more do I like the course that is all of the kind that we call "natural science."

It is surely unnecessary for me to say that I hold also for the full educational value of science that is applied. Science is science, whether devoted to the uses of life or whether it rests as its own reward; science does not need justification; no knowledge needs justification; it is for this reason that we should make no classification of "pure" and (by implication) "impure" science, any more than we should perpetuate the fiction of humanities and science: one is able to appreciate science for the sake of science at the same time that one applauds the application of it to medicine and agriculture. No person should ever attempt to apply scientific investigation until he understands and values science for its truth in the abstract.

We are misled by our phrases.

The speaker pointed out that the differences between the arts courses and the science courses are kept alive in part by the departmentalizing of our education, whereby each department of subject matter may become, at least in colleges and universities, a sort of an independent monarchy presided over by one king. In the larger institutions of higher learning, the great lines are separated into distinct colleges with separate and more or less autonomous administrative heads. These separations make the subjects to appear as if naturally differentiated and distinct, whereas in nature there are no such clear divisions. For administrative purposes, it may always be necessary, particularly with the growth of institutions, to separate the parts and to name them; but we should devise some way or system whereby the pedagogical aims can be brought together and one subject be brought to bear on the other. The departments in life are not as distinct as the departments in schools.

We need harmony in educational purpose rather than separation and antagonism. Whether or not we can make any change in the departmentalizing, we certainly can be careful not to suggest to the pupil that there are two camps, two realms with divergent aims, a superior and an inferior kind of subject-matter. We can also stop all the weary discussions of "culture"; the word really contributes nothing but confusion to education. The war has shown us how dangerous it may be, in another language form as a password; we should now be ashamed to use it.[9]

The second part of the address dealt with the place of science teaching in the development of personality and also of democratic ideas. It has recently been said that the teaching of science has resulted in the deterioration of character. So far as such evils have followed, it is not that science is inadequate to the highest results in human character, but rather that we have not yet learned how to use and to teach the vast treasures of fact and application that have overwhelmed us in recent times. Science is as capable of developing the higher moral and sentimental qualities as are the older subjects. We shall understand in due time that science is not merely a handmaiden to industry, but that it may expand the soul.

The speaker detailed some of the gains in intellectual poise and outlook that may come from a good teaching of natural science in the schools and higher institutions. He founded his discussion on the statement that the purpose of the quest of science is to find the fact and to know the truth. The truth is impartial, it invites a following to the logical conclusion; therefore, it trains directly in integrity of mind. The teaching of science stands always for the open mind. The man who prejudges or who starts with personal convictions does not become an investigator. He is more than likely to use the facts of science to uphold his own egotism; this is not science, however freely it may incorporate scientific facts into its processes. Science puts out no feelers to test public opinion. It is not dogmatic. It is not partisan, if its judgment is that of the open mind, seeking the truth. Undoubtedly very much of the spread of democracy in recent time is due directly to the teaching of science, whereby persons are taught to seek the fact before they draw their conclusions.

Science is never partial to any set of facts; it knows no "beliefs"; it is free to all men so far as they are able to understand; it is unselfish; it is adaptable to all persons, fitting their needs; in the quest of science there is no secrecy, no deals, no accommodations, no conspiracy, no favor, and no courtesy to high opinion that is not founded on rational investigation; the science method is not a secret method; it removes the fear of truth and the fear of dogma and the fear of nature. Science develops the individual, because every person makes his own investigation and takes nothing for

granted. It makes directly for the independence of the voter and for sta-
bility in public opinion. It is revolutionizing agriculture; no longer do we
plant in the moon and no more are we guided by the Babylonian signs. It
is time to introduce into politics the attitude of the open mind, independent of party programs, to approach public questions in something of the
spirit in which we approach the problems of science, desiring to know
the facts, the situation, and to decide after we know rather than before.
The facts of science are not discovered by debate or by argument. Neither
are we able to settle the tariff or any other public question by platform
polemics. We need first the facts, and these are to be obtained only as the
result of patient investigation by persons who are carefully trained and
have no theories to establish. Without the spirit of science permeating the
body politic, it is impossible to have a real democracy, for democracy is
not a form of government, nor is it freedom, but a state of society that
allows all citizens to partake and every one to develop his personality.
Democracy cannot be bestowed; it can only be achieved.

[From The Nature-Study Review, *vol. 14, no. 2, February 1918, pp. 43–47.]*

THE HUMANISTIC ELEMENT IN EDUCATION

L. H. BAILEY
President's Address at Annual Meeting

[The speaker explained that he had discussed "The Science Element in Education" before the Central Association of Science and Mathematics Teachers at Columbus, Ohio, November 30. He stated his point of view in that address: We are born to Things and to Phenomena. The regulated knowledge of Things and Phenomena is Science. As we depend on Things and Phenomena, so is the science of them essential; and what is essential is necessarily educational, if we are to live rationally. We are in error in supposing that there is a necessary educational line between "humanities" and "science," and we perpetuate error and hinder progress by the liberal use of these and other catch-words. We are misled by our phrases. In the present address the speaker sought still further to break down the prejudices between what may be called the old-line and the new-line subjects. A full abstract will be found in *School Science and Mathematics*, and an extract in *School and Society*, for December 29.]

We are born to People. Probably our first acquired knowledge is of father and mother. Human forms impress us so early that we never know that we never knew them. Brother, sister, family, the gradually enlarging circle of those of whom the child is "not afraid," make up the early experience. Soon the child begins to have consciousness of the many people, the strange people, those who quickly come and go, those on the street, in wagons, standing on the corners, waiting at the big places. The world is full of folks.

Soon the individuals begin to separate from the crowd. Faces become so familiar that the child names them and identifies them. Each one is unlike every other one. The child says that some persons are "funny."

Yet the moving crowd of human beings is the great fact of life. It is the great fact of the earth. These beings are gregarious. They move in long lines. They swarm in great masses. They colonize themselves in tense confusions that we call cities. Now and then one being separates itself and lives apart. That one is queer, clearly an aberrance. Most of us come back to the crowd as the meteor seeks the earth. Even when we are separate we talk in terms of the crowd. To go alone is unusual. When we go by ourselves we write a book about it.

What I mean to say is that human beings express habit and habitat, as do other animals. We are so accustomed to the habits that we think of them only to approve or to criticize. Yet essentially the habits of John Smith the Man are as interesting in themselves as are those of Lobo the Wolf, Black Beauty the Horse, or the Cat that Walked by its Wild Lone.[10] But we fail to observe John Smith objectively.

As there are laws of the Pack and laws of the Jungle, so are there laws of the Camp of Homo. At first the laws of the Pack and the Jungle and the Camp were probably much the same; but the Camp became crafty, self-willed, and it made weapons against the others. These weapons it turned also against the Other Camp. The Camp has come a long journey since then, but it has carried its weapons all the way.

The Camp found Speech and Handicraft. It found Importance, and set down its thoughts on stone and ivory and bones. It found Paper. Then it kept Records. Then did Literature begin. And in due time Men knew that they were Men, and wrote down the joy they had in thinking.

They thought about themselves and about Beings of another world; and so great and important were these Beings that man fashioned them in his own image and endowed them with his own qualities. So Man began to speculate, and to weave a vast web of fancy about himself and the Stars and the Things He Does Not Know. This web we call Literature, Philosophy, Art, Religion,—what you will.

And in due time Man came to be curious about the Things-Around-Him. He pried into them. He looked into crevices in rocks, ran his fingers along the seams of wood, found new metals, counted the eggs in a thousand nests, unravelled the flowers, searched for the alchemy, explored every wonder, enciphered the universe in formula and symbol. At some point in this long process he wrote down what he saw on papyrus or pieces of paper; then was Science born.

Very exact is Observation and very direct and true are Results. But these are first observations and first results. When we look again we begin to doubt. When we make a Conclusion we immediately set about to show that it is not true. They still say that there are "exact sciences;" if there are such, they must be those not founded on observation and experiment. I heard a man expound for an hour, with floods of numerals. He said that he had "proved" something. I do not know what it was.

So the deeper we settle into Science the more do we discuss and explain, which means only that we are trying dimly to understand. And the scientist becomes an hypothecist. To-day the plant-breeder is a mathematician, the zoologist is a speculator, and the geologist is a seer. And it endeth in Literature, Philosophy, Art, Religion,—what you will.

And it came to pass that men said one way was the best way and other men said their way was the best. And one man called his way Humanistic and the other called his way Scientific; and straightway they made much trouble for themselves.

One day we may forget distinctions that do not distinguish, and we may devote most of our energy to doing our piece of work well and to making ourselves to be as little children that we may teach simply and easily and directly.

Perhaps it would be impertinent, but I do not see how we can ever understand human beings or know what their habits mean or judge them fairly unless we observe them impartially and objectively. Now we judge them by ourselves. We think of them mostly as bearing "conduct" rather than as exhibiting characteristics. Never can we realize the brotherhood of man till we divest ourselves of prejudgment (which is prejudice), of assumed standards of ethics, and study human beings impersonally. Medicine could make no progress till it passed the idea of demons, of control by extra-terrestrial agencies, special providencies, and judgments for sin. Our actions and habits issue from causes and they follow courses which may be understood. We do not understand them by sitting in judgment, although by that means we may protect society. The new penology has its root here. We begin to see that conduct has a rational basis.

All the "humanities" in education are worth as much as the "sciences" in the training of the young, if there are as good teachers, with as good facilities, to teach the one as the other. All these subjects are organized out of the human mind; the same quest of truth is in them all; the same integrity of thought may characterize them all. It is not true that a subject is useful in education in proportion as it can be applied in the affairs of life. It is not true that any subject is even relatively useless because it cannot be "applied."

Man is as much a part of nature as is a pigeon or a trillium. Did not Huxley write on man's place in nature?[11] It is an incomplete nature-study that eliminates man from its range. What we now need above all else in nature-study is a good procedure on the observation of human beings.

If man is part in nature, if he has had a progressive evolution, then his habits and also his institutions are but parts of his natural history. Tradition itself is a phase of the natural history of the race, and becomes an essential part in any worth-while study of the race. These traditions express themselves as well in what we call science as in what we call classics. They are expressions of our development within our environment and in contact with our fellows. Against all this background, the discussion of the relative importance of the humanities and the sciences seems trivial and empty. These historic separations should now be forgotten, as against the common interests of mankind.

Always have I tried to present to you the wholeness of nature-study. From the first I have stood against the exclusive observation and study of the objects counted as "practical." This is not because I am opposed to the practical and the applied in education, but because such narrowing of the subject presents a wrong and restricted view of nature. In whatever the child takes up, I have wanted it to see the animal or the plant or the situation as a whole, and as part of its environment, and not merely as yielding certain products or benefits.

The interest in itself and its right to live,—this is the reason for the study of any living object, whether a frog, a cabbage, a horse, or a human being.

So should I be careful that nature-study does not degenerate into a study of attributes. In at least one State a law compels instruction in the elementary grades "in the humane treatment of animals and birds." The humane interest in "animals and birds" results naturally from a knowledge of them. The teaching of humane natural-history subjects as a detached and literary exercise is both weak education and insufficient morals. It is like teaching the odor of the rose.

It is the unfortunate impediment against nature-study, in the estimation of many persons, that it fits only partially into the regulated schemes of education so much prized at the present. Pressed into these patterns it loses much of its freedom. Situations in nature are unfortunately disregardful of a syllabus and unconcerned of "credits." Even our nature-study writers are likely to take the attitude that nature-study must be so regularized as to allow it to be handled uniformly in all schools by all teachers. We are verily obsessed of uniformity, as if it had merit in itself. By this dominated uniformity we withhold the best teachers, discourage the mutations that make for progress, and stand in the way of leadership. I think we should encourage departures.

It is possible, I am convinced, to apply enacted law to education for the purpose of safeguarding public funds and establishing an institution for the advancement of all the people at the same time that we allow the development of the full personality and initiative of strong teachers. Good system and method are much to be encouraged if they are in the nature of tested educational programs, founded on what we hope will some day be the science of education. This is very different from implanted governmental orders and insistence on the mere machinery of operation. Our law-made education, paper projects, and office regulation force our work into the plane of uniform mediocrity. All uniformity is mediocre.

I do not care to have nature-study similarly or equally taught in all schools. I hope something better for it than this. We are now in the grip of an artificial standardized system, matching well with the present theory of civilization. In due time, however, we shall return to the old conception of teaching, which is the principle of discipleship.

What, then, is my plea this morning? This only: that human beings are prime subjects for nature-study; that the old distinctions between the humanities and the sciences, represented in many catch-words, are essentially false; that nature-study stands for the spirit rather than for the form, and is to that extent a saving grace in the dominated systems of the day. I would make nature-study contribute to brotherhood. Nature is not an organized and classified procedure, as are the institutions of human affairs: the ultimate truth in nature is not yet discovered in statutory educational systems.

ACKNOWLEDGMENTS

For their generous financial support of this volume, which has enabled us to offer it in a free, open-access digital edition as well as in an affordable hardcopy edition, I extend my deep gratitude to Ann Habicht, Jack Padalino, the American Horticultural Society, the Antioch University Nature-Based Early Childhood Program, the Brandwein Institute, and #NatureForAll/International Union for Conservation of Nature Commission on Education and Communication. I am also particularly grateful to Cheryl Charles for her help marshalling many of these resources. Thanks also to all the supporters who donated to the book's online fundraising campaign. The generosity of these individuals and organizations made this volume much more accessible and affordable for the teachers for whom this book was always originally written, and I can't imagine a better way to launch the book and the Liberty Hyde Bailey Library into the twenty-first century. I offer my additional gratitude to the English Department at New York University, which has supported my work for years and which generously provided the funds necessary to hire an indexer for this volume.

The critical assistance and encouragement of many who make editions like this possible too often go unacknowledged. First, for her faith in this series and in this edition in particular, and for all her unflagging help shepherding this project through at the press during an unprecedented pandemic and a time of economic turmoil, I thank Kitty Liu. The iconic imprint of the Comstock Publishing Associates could not be in better hands. Thanks also to all the staff at Cornell University Press for their tireless work and patience, especially Martyn Beeny, Rebecca Brutus, Kristen Gregg, Eva Silverfine, Susan Specter, and Jacqulyn Teoh—it takes a village. I greatly appreciate the time and care that Lisa DeBoer took preparing the index. And to David Orr and Dilafruz Williams, for agreeing to contribute their wisdom to this volume, I am deeply and humbly grateful.

Many archivists and librarians went to great lengths to answer questions and make materials available electronically during the COVID-19 pandemic in which their institutions were closed to visitors. I thank especially Meredith Mann and the staff at the Brooke Russell Astor Reading Room for Rare Books and Manuscripts of the New York Public Library, which I was able to visit in November 2020 during a brief window in which they were able to reopen to researchers; Eisha Neely at Cornell University Library's Division of Rare and Manuscript Collections; Tim McRoberts and Jennie Rankin at Michigan State University (MSU) Archives and Historical Collections; Jim Ollgaard and Ed Appleyard at the Historical Association of South Haven, Michigan; Susan Smith at the Sumter County Library in Sumter, South Carolina; Andrew S. Russell at the Louise Pettus Archives and Special Collections at Winthrop University; Wade H. Dorsey at the South Carolina Department of Archives and History; Grace Pritchard-Woods at the Dean Close School in Cheltenham, United Kingdom; Lewis Wyman in the Manuscript Division of the Library of Congress; Adrienne Rusinko in Special Collections at Princeton University; and Flo Mauro at the Brandwein Institute and American Nature Study Society Archive. A particular note of thanks is due to my brother, Sam Linstrom, who followed up on a lead I had late in the production of this edition to access a cache of correspondence from and related to Julia Field-King at the Mary Baker Eddy Research Library in Boston; thanks also to the archivists there who helped him, Dorothy Rivera and Nathan Buchanan. My appreciation also goes out to the people at HathiTrust, which made many print resources still under copyright freely available online on a temporary basis during the pandemic. My bibliography attests to how essential that service became.

I owe a deep debt of gratitude to the late Jane L. Taylor, Founding Curator of the Michigan 4-H Children's Garden at MSU, who was one of my earliest cheerleaders in the world of Baileyana when I began working at the Liberty Hyde Bailey Museum in South Haven over a decade ago, and whose interest in Bailey's early teaching career while a college student in East Lansing led to the creation of an invaluable cache of documents in her archive at MSU. She passed away before this book was approved at the press, but the documentary trail she left, as well as her inspiring legacy in the field of outdoor learning, has informed this project from its inception.

Jane also volunteered to sit on the original advisory board for this series, where her guidance was deeply appreciated and today is sincerely missed, and alongside her I also thank the board's other members, Robert Dirig, Scott J. Peters, Daniel Wayne Rinn, Mary Swander, and Paul B. Thompson, for their continuing support and guidance as we seek to reintroduce and reappraise Bailey's vast corpus for the twenty-first century, beginning with the present volume. Each of these advisors has deepened my understanding of Bailey's significance, and each has also personally encouraged my scholarship in invaluable ways.

Thanks to the peer reviewers who helped hone the vision for this edition, its significance, and how it should be presented to new readers. Your feedback was generous and invaluable. Thanks also to Taylor Brorby, James R. Kates, A. G. Rud, and Paul B. Thompson, whose input on the potential of this book for course adoption helped to move the proposal along.

I am eager to express deep gratitude to my doctoral mentors, whose advice and patience allowed me to pursue this project alongside dissertation work: Una Chaudhuri, Sonya Posmentier, and Simón Trujillo. Thanks to Tom Augst and the Polonsky Foundation–New York University Digital Humanities Internship Program for providing me the time and tools to construct the website www.lhbaileyproject.com, a companion site to the Liberty Hyde Bailey Library. For their advice on certain sections of this manuscript, I am particularly indebted to Karen Penders St. Clair and Kathleen Allen, two passionate and erudite stewards of the story of the Cornell nature-study movement and its women leaders; to Patricia Crain, who leant her broad expertise in the American nineteenth century early in my research on this text to help me better understand the state of educational pedagogy when it was written; to Elaine Stephens and Jean Brown, who have always encouraged my work on Bailey and who lent their pedagogical expertise to my thinking early in this process; and to John A. Stempien, my perpetual coconspirator in all things Bailey whose willingness to bat ideas around at the drop of a text message has always kept my spirit up and energy flowing in the most obscure moments of research. I also thank the ever-generous community of Baileyphiles in and around Ithaca, who have been wonderfully generous on my various research pilgrimages there over the years, including Edward Cobb, William Crepet, Robert Dirig, Elaine Engst, Peter Fraissinet, Scott Peters, and Anna Stalter.

With all this help and support, the reader is yet sure to find errors or misrepresentations, and any such are mine alone.

To my parents, Robert and Rebecca, both master teachers in their own spheres, for modeling lives of learning and prioritizing ample time outdoors for my brothers and me when we were growing up, I will be offering thanks my whole life. My mother, a master teacher in early childhood education with over two decades of experience and two master's degrees, and the developer and de facto director of the Liberty Hyde Bailey Outdoor Learning Center at North Shore Elementary School in South Haven until her recent and well-earned retirement, has provided particularly invaluable feedback at several different stages of this manuscript's preparation, particularly in matters related to pedagogy and the practicalities of teaching children. It was a great privilege to be doubly guided through this process, both professionally and personally, by someone I so love and admire.

To the beaches and ravines of South Haven, where I learned so much about the world that I will never be able to translate into words, and where I absorbed perhaps some of the magic that Bailey also knew as he explored those same streams and dunes in the 1860s and 1870s, I owe much of who I am. To my brothers Ben and Sam, who accompanied me on those adventures, I remain truly grateful.

And to Monique, love of my life, for putting up with my penchant for too many projects and for encouraging and making possible these many pursuits, and for the unending gift of companionship and support, I cannot offer sufficient thanks. I am so grateful that little Chloe, who we scarce dreamed of when work on this book was begun and who nevertheless has made her entrance into the world well before it, will be lucky enough to grow up with such a loving momma. In some ways, all this, and everything before it, has really been preparation for her.

NOTES

Note that cross-references such as "see *Nature-Study Idea*, note 26" refer to sections in this volume.

Foreword

1. See also Jack, "Sower and Seer."
2. Comstock, *Handbook of Nature Study*, 16.
3. Comstock, *Handbook of Nature Study*, 22.
4. Bailey, *Holy Earth*, 21.
5. See Dewey, *Democracy and Education*; and *The Public and Its Problems*, 213. During a visit to the University of Vermont in his ninety-first year, Dewey read a passage from Bailey's *The Holy Earth*, suggesting synergy between Bailey's thought and that of Dewey (Rockefeller, *John Dewey*, 559).
6. Quoted in Lord, *Care of the Earth*, 201.
7. See Louv, *Last Child*, *Nature Principle*, *Our Wild Calling*, and "Outdoors for All." See also www.childrenandnature.org.
8. See Wu, *Attention Merchants*.
9. Bailey, *What Is Democracy?*, 45.
10. Bailey, *What Is Democracy?*, 106.
11. Bailey, *What Is Democracy?*, 105.
12. Bailey, *What Is Democracy?*, 114. In *The Holy Earth*, he wrote: "We do not yet know whether the race can permanently endure urban life, or whether it must be constantly renewed from the vitalities in the rear. We know that the farm and the back spaces have been the mother of the race. We know that the exigencies and frugalities of life in these backgrounds

beget men and women to be serious and steady and to know the value of every hour and of every coin that they earn" (28).

13. Bailey, *What Is Democracy?*, 123–135.

14. Bailey, *Universal Service*, 4, 18.

15. Bailey, *Universal Service*, 59–62, 163–164.

16. The hyphenated "agri-culture" is illuminated in Pretty, *Agri-Culture*.

Introduction

1. See, for instance, Intergovernmental Panel on Climate Change, *Climate Change and Land*, section 4.6 on "Impacts of land degradation on climate."

2. This is part of his thesis in *What Is Democracy?* (1918).

3. Morgan and Peters, "Foundations of Planetary Agrarianism"; Bailey, *Holy Earth*, 20–24, 25. Morgan and Peters apply the term "worldview transition" as it was developed by Thomas Berry.

4. Bailey, *Outlook to Nature*, rev. ed., 7.

5. See, e.g., Cruikshank, "American Herbartianism," 27–56.

6. Bailey, *Holy Earth*, 17–19.

Bringing Education to Life and Life to Education

1. See Williams, "Garden-Based Education."

2. See, for instance, Ardoin, Bowers, and Gaillard, "Environmental Education"; Kuo and Jordan, "The Natural World"; Kuo, Barnes, and Jordan, "Experiences with Nature"; and Williams and Dixon, "Impact of Garden-Based Learning."

3. See Wilson, *Biophilia*.

4. Kimmerer, "Traditional Ecological Knowledge," 433.

"It Is Spirit"

1. The epigraph from *The Tribune Farmer* appears here as it did in advertising copy for *The Nature-Study Idea*. See, for instance, Bailey, *Country Life-Movement*, 223.

2. "South Haven Is Home."

3. "South Haven Is Home."

4. "South Haven Is Home."

5. This is the story as Bailey told it in an undated transcript of an interview apparently conducted by E. Laurence Palmer and given over the WHCU radio station in Ithaca, NY. Liberty Hyde Bailey Papers, #21-2-3342, Division of Rare and Manuscript Collections, Cornell University Library, Ithaca, NY. He told an abbreviated version of the same story in an interview in South Haven in 1930, recorded in "South Haven Is Home." In the case of one line of dialogue, I have given the text as it appears in Louise Spieker Rankin's partial, unpublished biography of Bailey, housed in the Louise Spieker Rankin Papers at the Division of Rare and Manuscript Collections, Cornell University. Rankin's manuscript clearly provided the basis for the description of Bailey's early years to be found in Philip Dorf's 1956 biography, and Dorf notes that Rankin's work was "based on notes, letters, and some 30 interviews with Bailey during 1950–1951" (Dorf, *Liberty Hyde Bailey*, 250). Rankin's account matches Bailey's in the radio interview transcription very closely. As Dorf himself never met Bailey, I defer to Rankin here when the dialogue is not given in the Bailey transcription, and other details in the forgoing narrative also come from Rankin. See Rankin, unpublished biography, 3.4.8 and 3.4.18–21, and "South Haven Is Home"; cf. Dorf, *Liberty Hyde Bailey*, 20–21.

6. "South Haven Is Home."

7. She would recall being about thirty at the time of teaching Bailey in a 1909 letter to him, which would put the year around 1870, when Bailey would have been eleven or twelve. Julia Field-King to L. H. Bailey, September 14, 1909, Liberty Hyde Bailey Papers, #21–2–3342, Division of Rare and Manuscript Collections, Cornell University Library, Ithaca, NY. Rankin and Dorf both describe Field as "an Englishwoman," but, while she did eventually settle in England, her birthplace is listed in the Census for England and Wales of 1911 as "Montgomery, Illinois, U.S.A.," her age 71. The 1901 census corroborates her age and lists her birthplace as "America."

8. See Kohlstedt, *Teaching Children Science*, esp. 203–204.

9. Bailey, "Nature-Study Movement," 21, 24. This essay is available through the digital companion exhibition to this book at www.lhbaileyproject.com.

10. Palmer, "Nature-Study Philosophy," 40.

11. Rankin, unpublished partial biography, 1.6.14, Louise Spieker Rankin Papers, Division of Rare and Manuscript Collections, Cornell University. I owe my understanding of Bailey's early school teaching to the late Jane L. Taylor, whose research papers in the Michigan State University Archives and Historical Collections related to Bailey's time at the Carl School were invaluable.

12. Rodgers, *Liberty Hyde Bailey*, 25. The dates of Bailey's instruction at Carl School come from "Old Educators," 479. Dorf claims that Bailey "took over a grade school in a backwoods settlement" a year earlier than this, in the winter of 1878–1879, although it is not clear if this was Carl School or another district school in Michigan (39). That is the only description of Bailey's early teaching in Dorf's biography, and Rodgers does not mention it. In a late-life recorded interview with George H. M. Lawrence, Bailey says, "The first school I ever taught was up in central Michigan," and the transcribers of the audio recording, Frank Dennis and Jane Taylor, identify this as the Carl School (Bailey, interviews, 41).

13. "Old Educators," 481. All quotations and details about Bailey's teaching in this section come from this short article profiling Bailey through the memories of Donley. The article is housed in the Jane Taylor Collection, UA.17.292, Michigan State University Archives and Historical Collections, East Lansing, MI.

14. On his belief in education as foundational to social progress, see, for instance, his *Ground-Levels in Democracy*, esp. 7–27.

15. See *Nature-Study Idea*, note 26.

16. Comstock describes this origin for Cornell's nature-study work in her *Handbook of Nature Study*, xi–xii. See also the passage cut from Part I, Chapter VII of *The Nature-Study Idea*, included in this volume under Major Sections Restored from the First Edition.

17. Kohlstedt, *Teaching Children Science*, 266n40, 274n178.

18. Kohlstedt attributes the book's origins to Bailey's lectures; see *Teaching Children Science*, 281n96.

19. Including Charles Scott, at the influential Oswego Normal School where the "object method" was pioneered. See Kohlstedt, *Teaching Children Science*, 85.

20. In Bailey, *Gardener's Companion*, 22.

21. Bailey, *Onamanni*, iii.

22. L. H. Bailey to Mr. [George P.] Brett, September 20 and 28, 1899, Macmillan Company records, Manuscripts and Archives Division, the New York Public Library.

23. L. H. Bailey to Mr. Geo. P. Brett, January 3, 1901, Macmillan Company records, Manuscripts and Archives Division, the New York Public Library.

24. L. H. Bailey to Mr. George P. Brett, April 3, 1901, Macmillan Company records, Manuscripts and Archives Division, the New York Public Library.

25. Comstock, *Comstocks of Cornell*, 234.

26. Bailey, "Nature-Study Idea," 128.

27. Bailey, "Nature-Study Idea," 129.

28. See *Nature-Study Idea*, note 103.

29. Carl Fuldner has recently demonstrated that this "camera hunting" was promoted by many early photographer-naturalists as "an inclusive, participatory hobby" meant to "draw people out into the world" in order to experience nature firsthand ("Evolving Photography," 146). He ascribes the emergence of the genre of nature photography in the 1890s to this collective effort and notes the significance of Bailey's nature-study philosophy to the larger movement to encourage greater engagement with the natural world (150–151, 155–156). For a specific treatment of "The New Hunting" as it appeared in *Nature Portraits*, see Fuldner, "Evolving Photography," 200–201. Bailey was himself a significant nature photographer and an early pioneer of botanical photography; he was the subject of a major photobook manuscript by the great twentieth-century photography curator John Szarkowski titled *Liberty Hyde Bailey and the Survival of the Unlike*, which remains unpublished.

30. Bailey, *Nature Portraits*, 1.

31. Dorf, *Liberty Hyde Bailey*, 34. Cf. Rodgers, *Liberty Hyde Bailey*, 14.

32. Rankin, unpublished partial biography, 1.6.21–22.

33. See Comstock, *Comstocks of Cornell*, 228, 349, and 386.

34. For a recent overview of these women, see Allen, "Women of Cornell"; and St. Clair, "Finding Anna," 34–85.

35. Comstock, *Comstocks of Cornell*, 368.

36. In a 1907 letter from Bailey to Sarah Payne, quoted in Kohlstedt, *Teaching Children Science*, 174. Bailey's reassurance would have been appropriate for many nature-study teachers at the time, who appear to have been generally more effective than a reader of the contemporary scientific literature might be led to believe; analysis of written records of actual nature-study instruction during this period demonstrates that, despite the criticisms of the scientific establishment, teachers felt that they could effectively incorporate the kinds of nature-study activities suggested by groups like the Cornell Bureau of Nature-Study and that such work was a distinct benefit to their classrooms. Contrary to twentieth-century historiography, teachers were not confused by or ill-equipped to implement nature-study resources. See Doris, "Practice of Nature-Study."

37. Beal's two essays are reproduced in this volume under Related Writings.

38. See "It Is Spirit," note 18.

39. Bailey, *Universal Service*, 16.

40. Bailey, *Holy Earth*, 17–19, 25, 12–13.

41. Bailey, [Ninetieth Birthday Speech], 25–26.

42. Bailey, *Holy Earth*, 20–24.

43. Bailey, *Country Life-Movement*, 220.

44. Late in his life, the accomplished Black landscape architect David Williston wrote in a letter to Cornell University that Bailey "inspired me to study plant life, and I am happy to state that I was a student of his for more than 60 years." Quoted in Kammen, *Part & Apart*, 33–34.

45. Roscoe Conkling Bruce to Liberty Hyde Bailey, 1903, Liberty Hyde Bailey Papers, #21-2-3342, Division of Rare and Manuscript Collections, Cornell University Library, Ithaca, NY.

46. Washington, *Working with the Hands*, 159.

47. Washington, "Negro Farmer."

48. Armitage, *Nature Study Movement*, 189–191.

49. For a good recent history situating Washington and Carver in the context of the struggle for land sovereignty and Black liberation, see Monica White, *Freedom Farmers*, esp. 3–62. For more on the Black nature-study movement in the South, see Glave, *Rooted in the Earth*, 97–114.

50. Quoted in Kohlstedt, *Teaching Children Science*, 127.

51. Quoted in Kohlstedt, *Teaching Children Science*, 127, 86.

52. L. H. Bailey to Mr. George P. Brett, October 26, 1908; L. H. Bailey to the Macmillan Company, October 17, 1908; and the Macmillan Company to Professor Bailey, October 21, 1908, Macmillan Company records, Manuscripts and Archives Division, the New York Public Library.

53. L. H. Bailey to Mr. [George P.] Brett, December 16, 1907, Macmillan Company records, Manuscripts and Archives Division, the New York Public Library.

54. Peters and Morgan, "Country Life Commission," 294.

55. L. H. Bailey to Mr. E. C. [*sic*] Brett, May 14, 1909, Macmillan Company records, Manuscripts and Archives Division, the New York Public Library.

56. L. H. Bailey to Mr. [George P.] Brett, May 21, 1909, Macmillan Company records, Manuscripts and Archives Division, the New York Public Library.

57. L. H. Bailey to Mr. [George P.] Brett, May 25, 1909, Macmillan Company records, Manuscripts and Archives Division, the New York Public Library.

58. See https://www.civicecology.org/. On the program's imitation across North America, see Kohlstedt, *Teaching Children Science*, 78.

59. Quoted in Kohlstedt, *Teaching Children Science*, 93, 269n87 (caps in original).

60. Armitage, *Nature Study Movement*, 209. Also see Kohlstedt, *Teaching Children Science*, 233–236.

61. Kohlstedt, *Teaching Children Science*, 91–92.

62. Kohlstedt, *Teaching Children Science*, 237, 192–199.

63. Flo Mauro, email message to author, August 13, 2020. Quotation, and more on the organization's active history and publications, in Brandwein Institute, "Brief History."

64. Kohlstedt, *Teaching Children Science*, 199.

65. Linstrom, "Land, Labor, Literature," 117–118.

66. Julia Field-King to L. H. Bailey, September 14, 1909, Liberty Hyde Bailey Papers, #21-2-3342, Division of Rare and Manuscript Collections, Cornell University Library, Ithaca, NY. In addition to that letter's heading, Field-King is listed as living at Longleat on the Census of England and Wales for 1911, listed as a "friend" living with Mary Ann Woolgar, head of household, both 71 years old and widowed, and Constance Myley, "servant," 28 years old.

67. Julia Field-King, "Application for Study at Massachusetts Metaphysical College, with Notations," January 28, 1888, item number L19194. Mary Baker Eddy Papers, Mary Baker Eddy Library, Boston, MA.

68. Cook County, Illinois, Marriages Index, 1871–1920. Ancestry.com, https://www.ancestrylibrary.com/discoveryui-content/view/1078047:2556?tid=&pid=&queryId=ee189de59012a8956e75ba08af81ba36&_phsrc=lUp1&_phstart=successSource.

69. Fine, "Medical Education."

70. Julia Field-King, "Application for Study at Massachusetts Metaphysical College, with Notations," 28 Jan. 1888, item number L19194. Mary Baker Eddy Papers, Mary Baker Eddy Library, Boston, MA.

71. The pages of handwritten correspondence between Eddy and Field-King in the Mary Baker Eddy Papers number well into the hundreds.

72. *Charges vs. Mrs. Julia Field-King, C.S.D. of England, Jan.-Feb., 1902*, item number 214c.35.059, Mary Baker Eddy Papers, Mary Baker Eddy Library, Boston, MA.

73. *Charges vs. Mrs. Julia Field-King*. In a letter to an unnamed "friend" later that year, Field-King writes, "You see I am not charged with any sin against God; only with violating some of the many, many rules of the modern Leviticus, called the Church Manual. What a destroyer of spontaneous love and gratitude, and of honest demonstration as the test of a true Christian Scientist it is." Julia Field-King to Friend, May 8, 1902, Mary Baker Eddy Papers, Mary Baker Eddy Library, Boston, MA.

Note on the Text

1. He did this, for instance, between the 1915 and 1916 printings of *The Holy Earth*; see my editorial introduction to that volume.

The Nature-Study Idea

1. The "leading technical journal" was *Science*, and the "contributor" was Bailey's former professor and colleague at Michigan's State Agricultural College (now Michigan State University), William J. Beal. Bailey's disagreement with Beal's method of evaluating nature-study is notable partly because Beal had recruited Bailey and mentored him through college, later recommending him to work in the herbarium of leading botanist Asa Gray at Harvard (see *Nature-Study Idea*, note 42) and effectively launching Bailey's scientific career. See "It Is Spirit," in this volume, as well as the Related Writings section, for more on their disagreement. *The Nature-Study Idea* responds to the accusations put forward by Beal by insisting that nature-study is a movement of the common schools to put children into sympathy with nature with the end goal of greater happiness in life, not to teach science for science's sake. For another recent consideration of the Bailey/Beal debate, see Schulze, *Degenerate Muse*, chapter 1, esp. 59–61.

2. The "common schools," a term coined by the American educational reformer Horace Mann (1796–1859), were the predecessors to today's public schools, intended to provide publicly funded education at no cost to students.

3. *The Nature-Study Idea* went through at least four distinct editions during Bailey's lifetime, two with Doubleday (1903 and 1905) and two with Macmillan (1909 and 1911), in addition to numerous reprints. The third, 1909, edition, for which Bailey wrote the remainder of this chapter, incorporated thorough and substantial revisions and additions. The present text is based on the 1920 printing of the fourth (1911) edition. For more on Bailey's relationships with these publishers and the book's various editions, see "It Is Spirit" and the Note on the Text, this volume.

4. Bailey served as the Dean of the College of Agriculture at Cornell University from 1903 to 1913, which in 1904 by state legislative act became the New York State College of Agriculture and which grew rapidly under Bailey's administration. He retired from administration and teaching in 1913 at the age of fifty-five after an exhausting decade that some thought might lead to a nervous breakdown. See Colman, *Education & Agriculture*, 157–250, and Ethel Z. Bailey, interview, 25–30.

5. Throughout the nineteenth and into the twentieth century, much of early childhood education in the United States consisted of recitation—the student reading from a book or reciting from memory to the class, and the teacher drilling the student on the quality of elocution as well as the content recited. "Speech-education" here may refer to a combination of recitation and other practices like rhetorical or elocution training, which prepared students to be able to give speeches. The Protestant Reformation was a movement sparked by the German monk Martin Luther in sixteenth-century Europe. Luther famously translated the Bible into German so that common people could read it directly, rather than rely upon the interpretation

of priests in worship services, which, along with the advent of the printing press, led to an out-pouring of religious publications and an increase in literacy. My thanks to Patricia Crain for her help thinking through Bailey's "speech-education" comment.

6. The nature-study program at Cornell University was begun by an 1894 state appro-priation, thirteen years before this section of the chapter was first published in the book's third edition, under the initial leadership of Isaac P. Roberts, who turned it over to Bailey the following year (Comstock, *Handbook of Nature Study*, xi–xii). The program grew rap-idly under Bailey's administration, during which time he hired Anna Botsford Comstock, who along with Bailey would become a major leader of the nationwide nature-study movement (on Comstock, see *Nature-Study Idea*, note 128). Comstock described Bailey as "the inspir-ing leader of the [Cornell nature-study] movement, as well as the official head" (*Handbook of Nature Study*, xii), although this seems modest on her part: she certainly shared the role of "inspiring leader."

7. Bailey could indeed claim personal experience with such teaching, having himself taught in a rural schoolhouse called the Carl School near East Lansing to help support his un-dergraduate studies, in addition to his extensive work with teachers in New York State and with the nature-study faculty at Cornell. On his teaching at the Carl School, see "It Is Spirit," this volume, as well as "Old Educators," 79–82, and Ethel Z. Bailey, interview, 83–84. For another perspective on Bailey's wariness about pedagogical theory and the psychology of ed-ucation, see *Nature-Study Idea*, note 114.

8. This strategy of preparing teachers with a pedagogical approach and outlook rather than through "an outline for class work" characterizes much of what Bailey and his Cornell colleagues promoted in their nature-study publications, but that doesn't mean Bailey didn't publish more practical textbooks. Even before the first edition of *The Nature-Study Idea*, he had published *Lessons with Plants* (1897), written for use by teachers, and *Botany: An El-ementary Text for Schools* (1900), written for direct student use. Both provided content to work from and suggestions for activities, but they avoided dictating class work or how the teacher should use them and encouraged spontaneous outdoor exploration. In 1908, just one year before this section of his third-edition text was published, Bailey published another bo-tanical textbook, geared more clearly toward younger pupils, titled *Beginners' Botany*, and in 1913 he would lightly revise his elementary textbook of 1900 under the new title *Botany for Secondary Schools* (it seems to have been deemed too advanced for elementary students after all). Also in 1908, the text of *Beginners' Botany* would simultaneously appear as the first part, titled "Plant Biology," of the larger textbook *First Course in Biology*, cowritten with Walter M. Coleman, who contributed the book's second and third parts on "Animal Biology" and "Human Biology." Apparently, in the 1910s and 1920s, *Beginners' Botany* was variously re-worked for use specifically in Macmillan's Canadian School Series, including at least three re-gionally specific versions: one for Ontario, one for Manitoba and Saskatchewan, and another for Nova Scotia and New Brunswick. Other versions may also exist, but I have not been able to locate them. It seems to have been a successful text.

9. Portions of this and the following chapter also appeared in a slightly different form one month after the first edition of *The Nature-Study Idea* was published, as an article in the May 1903 issue of *Our Day* titled "The Nature-Study Idea: Being an Account of How the Term Originated and What It Really Means."

10. Jean Louis Rodolphe Agassiz (1807–1873), known as Louis Agassiz, the famous Swiss scientist who became professor of biology and geology at Harvard University, head of its Lawrence Scientific School, and founder of its Museum of Comparative Zoology. Among his students was the botanist William J. Beal, Bailey's undergraduate mentor whose critique of Bailey's nature-study philosophy motivated Bailey to publish this book (see the essay "It Is Spirit," this volume, and *Nature-Study Idea*, note 1). In the summer of 1873, Agassiz opened the Anderson School of Natural History on Penikese Island, which was the first professional

biological field station established in the United States and was intended specifically to train teachers, women as well as men, in Agassiz's method of teaching natural history through direct contact with nature. "Study nature, not books," was inscribed on a plaque at the entrance to the Penikese laboratory and would become a motto of the nature-study movement. Agassiz died in December of 1873, and the school lasted only one more summer, but many of the forty-four teachers who studied under him at Penikese became leaders in the nature-study movement in later decades. See Armitage, *Nature Study Movement*, esp. 14–22, Kohlstedt, *Teaching Children Science*, esp. 11–36, and Lurie, *Louis Agassiz*.

However, like many early nature-study theorists who came after him, Agassiz was also a believer in the since-debunked eugenic theory of recapitulation in child development, the idea that human development from fetus to adult strictly mimicked species evolution and that different races of humans evolved over time in a similarly correlated process, resulting in different racial groups representing different stages of human development. For more on Agassiz's racism, see Menand, *Metaphysical Club*, 103–116. "Although few nature study teachers drew the same overtly racist conclusions that Agassiz did, many thought the theory [of recapitulation] provided scientific support for the idea that children learned through contact with nature because, developmentally, they were in a savage state. For nature study advocates, learning from nature was simply another way of describing the thorough grounding in basic natural history they aimed to impart to ready students. The theory of recapitulation 'proved' that children had natural affinities for basic scientific exploration" (Armitage, *Nature Study Movement*, 73–74). For more on eugenics in the nature-study movement, see Armitage, *Nature Study Movement*, 71–91, as well as *Nature-Study Idea*, note 114, and "It Is Spirit," this volume.

11. See *Nature-Study Idea*, note 6.

12. Clifton F. Hodge (1859–1949), American professor of physiology at Clark University and author of *Nature Study and Life* (1902), which was cited in Beal's article criticizing Bailey's nature-study writings for being unscientific (see "It Is Spirit" and Related Writings). Hodge believed that depriving children of the opportunity to raise a plant led to a greater rate of crime in adulthood, that every city should contain a game preserve, and that people should plant trees on their properties and control their cats in order to benefit bird populations (Armitage, *Nature Study Movement*, 123). He was notoriously acerbic in his attacks both on scientists who dismissed nature-study and on nature-study proponents with whom he disagreed, and Bailey once wrote to John G. Coulter that, while he admired Hodge, "he cannot separate his work from personalities and he handicaps his efforts thereby" (quoted in Kohlstedt, *Teaching Children Science*, 299n82). Hodge critiqued Bailey's attention to detail in nature-study pedagogy as a "knot hole method" beginning with "a broom splint, a sliver of pine or a knot hole" to fill time. He was influenced by the recapitulation theory advocated by Hall (Kohlstedt, *Teaching Children Science*, 189; on recapitulation, see *Nature-Study Idea*, notes 10 and 114).

G. Stanley Hall (1846–1924), American psychologist, educator, and president of Clark University who studied under pragmatist philosopher William James and among whose students was John Dewey (see *Nature-Study Idea*, notes 19–20). Hall was a prominent eugenicist who, like Agassiz, believed in recapitulation theory and thought that human evolution was mimicked by individual human development from fetus to child to adolescent to adult (see "It Is Spirit" and *Nature-Study Idea*, notes 10 and 114). Hall also argued for a more "scientific" nature-study, and he believed men more capable of such instruction than women; in his introduction to Hodge's *Nature Study and Life*, he cautioned against the trend of more "sentimental" nature-study that he associated with "effeminization" (xv). For more on Hall in the context of sexism in nature-study, see Kohlstedt, *Teaching Children Science*, 168–172, and for Hall's biography see Ross, *G. Stanley Hall*. It may be significant that Bailey gives only one sentence to these two influential thinkers, claiming that Cornell's nature-study work had begun two years before Hodge's at Clark—on Bailey's role in empowering women within the nature-study movement, see the essay "It Is Spirit," this volume.

13. Socrates (ca. 470–399 BCE), Greek moral philosopher credited as a founder of Western philosophy.

Aristotle (384–322 BCE), Greek philosopher, polymath, and founder of the Lyceum, who studied under Socrates's student Plato and pioneered the systematic and empirical study of nature. Agassiz (*Nature-Study Idea*, note 10) wrote of Aristotle, "The great mind of Greece in his day, and a leader in all the intellectual culture of his time, he was especially a naturalist" ("Natural History," 1).

John Amos Comenius (1592–1670), Moravian monk, philosopher, and pedagogue considered the father of modern or progressive education. An inspiration to nineteenth-century nature-study advocates, Comenius wrote in his *Physics*, "Why, say I, should we not, instead of these dead books, lay open the living book of Nature, in which there is much more to contemplate than any one person can ever relate and the contemplation of which brings much more of pleasure, as well as of profit?" (quoted in Armitage, *Nature Study Movement*, 47).

Johann Heinrich Pestalozzi (1746–1827), Swiss educational reformer whose pedagogical theories, set out in the novels *Leonard and Gertrude* (1781) and *How Gertrude Teaches Her Children* (1801), influenced the development of "object teaching," in which children learn from observing and interacting with physical objects rather than from books and abstract concepts, and which in turn influenced the development of nature-study (Armitage, *Nature Study Movement*, 48–49, 22; Kohlstedt, *Teaching Children Science*, 112). Pestalozzi's ideas were brought to American education by the geologist and utopian founder of the New Harmony commune, William Maclure (1763–1840), and the educational reformer, Horace Mann (Kohlstedt, *Teaching Children Science*, 14, 27; on Mann, see *Nature-Study Idea*, note 2). The Pestalozzian method was concretized as the "object method" at the Oswego Normal School (Kohlstedt, *Teaching Children Science*, 30), which Bailey discusses later in this chapter.

Jean-Jacques Rousseau (1712–1778), Genevan social philosopher whose novel and treatise *Emile, or On Education* (1762) deeply influenced the pedagogical theory of Pestalozzi (Armitage, *Nature Study Movement*, 47–48).

Friedrich Wilhelm August Fröbel, or Froebel (1782–1852), German pedagogue and student of Pestalozzi who innovated the "kindergarten" idea and coined the term. Froebel was another influential continental pedagogue for American nature-study advocates, arguing that exposure to nature helped inculcate moral lessons in children and that teachers should "bring their personal educational experiences and theoretical training to the study of the natural environment in and around their schools" (Kohlstedt, *Teaching Children Science*, 112) and give children space to generate their own activities (Armitage, *Nature Study Movement*, 49; Kohlstedt, *Teaching Children Science*, 30, 126).

14. Richard Keller Piez (ca.1865–1946), professor of manual training, drawing, and psychology at the Oswego Normal School from 1893–1937 ("Dr. Piez, 81, Dead"; *History of the First Half*, 87–88). Piez believed that teachers should be active in the larger community in which the school is located, and he helped establish public playgrounds in Oswego (*History of the First Half*, 87–88).

15. Piez apparently refers to recitation (see *Nature-Study Idea*, note 5). Object teaching sought to correct this "mechanical" method of rote memorization by allowing children to learn through experiential engagement, or what John Dewey popularly described as "learning by doing."

16. Alpheus Hyatt (1838–1902), American biologist and paleontologist, curator of the Boston Society of Natural History, and professor at the Massachusetts Institute of Technology (1870–1877) and Boston University (1887–1902).

Lucretia Crocker (1829–1886), science educator, education instructor at the Massachusetts State Normal School (1850–1854), professor of mathematics and astronomy at Antioch College (1857–1859), and founder of the Women's Education Association (WEA) in 1872.

Hyatt and Crocker both studied under Agassiz, Crocker at the Penikese school (Brooks, "Alpheus Hyatt," 313–315; Kohlstedt, *Teaching Children Science*, 20, 43; on the Penikese school, see *Nature-Study Idea*, note 10). With the support of the WEA, Hyatt and Crocker coordinated school programs with the museum of the Boston Society of Natural History, and in the 1870s they collaborated on the Teacher's School of Science, which provided Saturday and evening classes to teachers in the Boston area and where Crocker came to hold the title "Nature Study Supervisor." Crocker is remembered as "among the first women who moved into school administration" (Kohlstedt, *Teaching Children Science*, 25).

17. Edward Austin Sheldon (1823–1897), American educator, administrator, and founding president of the Oswego Primary Teachers' Training School, later known as the Oswego Normal School and today the State University of New York at Oswego. In the 1860s, he began to implement the Pestalozzian use of objects for teaching children in what became known as "object lessons," also called the Oswego method. The school also pioneered the supervision of student teachers as a means of educational instruction. Sheldon's groundbreaking *Lessons on Objects* (1863) identified his name with the pedagogical movement and influenced the thinking of Colonel Francis W. Parker, whom Bailey also mentions in this paragraph (Kohlstedt, *Teaching Children Science*, 30, 250n100, 42; on Parker, see *Nature-Study Idea*, note 19). Sheldon's autobiography, edited by his daughter Mary Sheldon Barnes, was published in 1911.

18. Henry Harrison Straight (1846–1886), American geologist and professor who studied under Agassiz at the Penikese school (Armitage, *Nature Study Movement*, 23; Kohlstedt, *Teaching Children Science*, 43; on Agassiz and the Penikese school, see *Nature-Study Idea*, note 10). Straight pursued advanced studies at Cornell before joining Oswego, and there he became an admired colleague of Anna Botsford Comstock, a prominent leader of the nature-study program at Cornell (see *Nature-Study Idea*, note 128), who described "Professor and Mrs. Straight" as "very superior people" (Comstock, *Comstocks of Cornell*, 113; Kohlstedt, *Teaching Children Science*, 254n34). As Bailey describes in this paragraph, Straight helped move object teaching towards nature-study through "correlation" of subjects and the introduction of living objects. He advocated for schools to teach to the "complete life" of their students, and he integrated field trips to encourage experiential learning about the natural world. His work was cut short by tuberculosis (Armitage, *Nature Study Movement*, 23–24; Kohlstedt, *Teaching Children Science*, 43).

Nathaniel Southgate Shaler (1841–1906), American paleontologist, geologist, and conservationist who studied under Agassiz and became a professor at Harvard for many years. He first suggested to Agassiz the idea of the summer school at Penikese and taught on the school's faculty (Armitage, *Nature Study Movement*, 16–17, 54; Kohlstedt, *Teaching Children Science*, 249n73). Shaler also led his own part-time courses in continuing education at Harvard (Kohlstedt, *Teaching Children Science*, 41). While Shaler would eventually depart from Agassiz by accepting a neo-Lamarckian version of Darwinian evolution, he held onto the scientific racism of his mentor, falsely believing humans to be divisible into separate racial species, and he was an apologist for southern slavery through most of his life (Livingstone, *Nathaniel Southgate Shaler*, 124–125, 138–143).

19. Colonel Francis Wayland Parker (1837–1902), American Civil War veteran, educational reformer, and founding principal of the Cook County Normal School in Chicago, whom John Dewey dubbed the "father of the progressive education movement" (Armitage, *Nature Study Movement*, 71; Kohlstedt, *Teaching Children Science*, 29). Inspired by Edwin Sheldon's writings (see *Nature-Study Idea*, note 17), he traveled to Europe in the 1870s and spent over two years studying the pedagogical methods developed by Pestalozzi and Froebel

(both by then deceased) firsthand. He began to put those ideas to practice first as superintendent of schools in Quincy, MA, and then as principal at Cook County in 1883, where he hired Henry Straight that year (see *Nature-Study Idea*, note 18) and Wilbur Jackman in 1889 (see *Nature-Study Idea*, note 20). Cook County Normal School became its neighborhood's public school shortly after Parker arrived, and it quickly gained a reputation for educational excellence and an innovative curriculum built around experiential learning. When Dewey joined the faculty of the University of Chicago, he enrolled his children at Cook County Normal before starting his famous Laboratory School at the university.

In 1898, philanthropist Anita McCormick Blaine hired Parker as president of her newly founded Chicago Institute, a private model school based on the "new education" philosophy based around nature-study, and Jackman was hired as dean and head of the high school. The institute never reached self-sufficiency, and by 1901 it merged with the Laboratory School to form the University of Chicago's School of Education, with Parker as the Director of the School of Education (overseeing the practice school) and Dewey as a professor in the Department of Education. The program struggled after Parker's death the next year, and tensions flared between Dewey and Jackman, as under Dewey's directorship courses in educational theory increased while nature-study decreased (Armitage, *Nature Study Movement*, 52–53, 55–56; Kohlstedt, *Teaching Children Science*, 29, 42–57). Kohlstedt notes that a "significant part of Parker's success was, as an admirer put it, his 'faith in teachers, and this belief inspired them to accomplishment. It created in them a power which had not previously existed.' His positive assumptions about their skill and motivation appealed to teachers who were too often reminded of their inadequacies, both by advertisers who sold them journal subscriptions and textbooks and by administrators who kept expanding their responsibilities" (*Teaching Children Science*, 43).

20. Wilbur Samuel Jackman (1855–1907), American educator and author of *Nature Study for the Common Schools* (1891) and *Nature Study for Grammar Grades* (1899). Jackman grew up on a farm in Pennsylvania, and, prior to the period Bailey describes, he had studied under Nathaniel Shaler (see *Nature-Study Idea*, note 18) at Harvard. Bailey's neglect to describe Jackman's move to the Chicago Institute and then the School of Education at the University of Chicago may reflect Jackman's own wishes and dissatisfaction with the direction that the School of Education took after Parker's death (see *Nature-Study Idea*, note 19). The rift that grew between Jackman and Dewey in that context may help explain the fact that Dewey receives no mention in Bailey's book despite his towering place in the history of progressive education (on that rift, see Kohlstedt, *Teaching Children Science*, 196). Dewey's theories, nevertheless, bear striking resemblances with those articulated by Bailey and the Cornell nature-study leaders, and in books like *The School and Society* (1899) and *Democracy and Education* (1916), Dewey emphasized the importance of education through experiential learning as the foundation for a democratic society (Westbrook, *John Dewey*, 23–26, 104–111). By the time Dewey left to join the faculty of Columbia University in 1904, Jackman had stepped down from the position of dean of the Laboratory School to become principal of the university's elementary school and edit the journal *Elementary School Teacher* (Armitage, *Nature Study Movement*, 53–55; Kohlstedt, *Teaching Children Science*, 51–56). His impact on nature-study in the 1880s and 1890s, however, was immense, as Bailey makes clear in this paragraph, and his "five years' connection" with Pittsburgh High School supports Bailey's thesis that the nature-study movement emerged organically from teachers in the common, or public, schools (see Part I, Chapter I). Cornell nature-study professor Anna Botsford Comstock (see *Nature-Study Idea*, note 128) taught alongside Jackman in August 1902 for the State Teachers' Institute in Harrison, Ohio, and she described Jackman as "the father of Nature Study in America, and a man of high ideals and great accomplishment" who "believed that all elementary education should have as its foundation, Nature Study" (*Comstocks of Cornell*, 237).

21. Arthur Clarke Boyden (1852–1933), later principal of the State Normal School at Bridgewater (1906–1932) and, when it became Bridgewater State Teachers College, its first president (1932–1933), and author of *Nature Study by Months: Part I., for Elementary Grades* (1898). Boyden's father, Albert G. Boyden, served as principal of the school from 1860 to 1906, during which time Arthur Boyden attended the Normal School and graduated in 1871. After graduating from Amherst College in 1876, the younger Boyden taught mathematics at Chauncey Hall School in Boston for three years before joining the faculty at Bridgewater under his father. In addition to the work outlined by Bailey here, Arthur Boyden sat on the science subcommittee of the influential "Committee of Ten" that was appointed in 1892 by the National Education Association to standardize American school curricula. Among that subcommittee's recommendations was the widespread adoption of nature-study for early childhood education (Boyden, *History of Bridgewater*, 73, 81–82). That subcommittee's use of the term "nature-study" in its report helped circulate and validate the work that Wilbur Jackman had been promoting in his textbooks (Kohlstedt, *Teaching Children Science*, 47–49; on Jackman, see *Nature-Study Idea*, note 20). Before becoming principal, Arthur Boyden also organized a Teachers' School of Science consisting partly of former students of Alpheus Hyatt (see *Nature-Study Idea*, note 16) in 1891, an association promoting children's gardening in Bridgewater, and a two-acre Natural Science Garden at the normal school in 1907 (Boyden, *History of Bridgewater*, 83–86). (Conflating the names of the two Boydens, Kohlstedt claims that "Arthur G. Boyden" studied under Agassiz [*Teaching Children Science*, 20]; since Agassiz died in 1873, this most likely refers to Albert Gardner Boyden, who did contribute to the science-teaching philosophy that would inspire Arthur Clarke Boyden's nature-study work. On the elder Boyden, see *Albert Gardner Boyden and the Bridgewater State Normal School*, written by Arthur Clarke Boyden, 1919.)

22. George H. Martin (1841–1917), teacher at the State Normal School in Bridgewater, administrator at the state and local levels in the public schools of Massachusetts, and author of *A Text Book on Civil Government in the United States* (1875) and *The Evolution of the Massachusetts Public School System* (1894). Martin had studied at the Bridgewater Normal School under Principal Albert Gardner Boyden (see *Nature-Study Idea*, note 21), graduating in 1863. After many years teaching, he became involved in a local board of education in the 1880s and 1890s and was appointed secretary to the Massachusetts State Board of Education in 1905, serving in various capacities on the state board until 1911. In his later years, he worked closely with his Bridgewater colleague and Albert Boyden's son, Arthur Clarke Boyden (see *Nature-Study Idea*, note 21), who described Martin as "one of the earliest advocates of industrial education" (Boyden, *History of Bridgewater*, 72, 46, 71, 87, 114). Bailey avoids the term "industrial education" in *The Nature-Study Idea*, but two years later, in *The Outlook to Nature* (1905), he writes:

All this constitutes the new "industrial education,"—an education that uses the native objects and affairs of the community as means of training in scholarship, setting the youth right toward life, making him to feel that schooling is as indigenous and natural as any other part of his life, that he cannot afford to neglect schooling any more than he can neglect the learning of a business or occupation, that schooling will aid him directly in his occupation, that the home and school and daily work are only different phases of his own normal development, and that common duties may be made worthy of his ideals. Unfortunately, the term 'industrial education' is ordinarily understood to mean direct training for the trades; therefore it would be a great gain to a clear understanding of the subject if some other term could be used for this new and pedagogically sound idea. [. . .] In my own mind, the term "nature-study" is large enough, for I think of "nature," in this relation, as expressing the natural method of education, whereby the pupil is educated at first in the terms of the world he lives in; but the term has been so long used with another signification that it cannot be pressed into service for the larger and fuller idea. [. . .] For the time being, therefore, I see no better term than industrial education, with the reservation that it mean much more than commercial education [. . .]. (181–183)

23. Amos Markham Kellogg (1832–1914), superintendent of the experimental department of the New York State Normal School at Albany (Kellogg, *School Management*, i), later editor of *The School Journal* and *The Teacher's Institute* and author of books including *School Management: A Practical Guide for the Teacher in the School-Room*; *Pestalozzi: His Educational Work and Principles*; *Elementary Psychology*; and a number of short books of pedagogy and practical classroom exercises (Kohlstedt 254n25). In his editorial work and writing he strenuously promoted the ideas of Francis Parker ("Mr. Amos M. Kellogg," 380; on Parker, see *Nature-Study Idea*, note 19). *The School Journal* had taken a national focus and had not been titled *New York School Journal* for many years by the time of *The Nature-Study Idea*'s publication.

24. Frank Owen Payne (1859–1922), teacher in Corry, PA, and Chatham, NJ (Minton, "Nature-Study Movement," 84), and author of books including *Geographical Nature Studies* (1898) and *Manual of Experimental Botany* (1912) as well as *One Hundred Lessons in Nature Study around My School* (1895). He later became the first principal of the high school at Glen Cove, NY, and in 1913 took a position at the High School of Commerce in New York City, where he served as chairman of biology from 1916 until his death in 1922. In his later years, he was better known as a critic of American sculpture ("Frank O. Payne," 442). Minton notes Payne's debt to Comenius and claims that Payne first used the term "nature-study" ("Nature-Study Movement," 103–104).

25. An editorial in the third issue of *The Nature-Study Review* defends the hyphen this way: "similar compounds, such as nature-worship and nature-print are in the leading dictionaries hyphenated [. . ., and] there is a special argument for the hyphenated form in that we now have reason to speak of biological and physical nature-study, and the hyphen makes it clear that the adjective modifies the combined words" ("Nature Study or Nature-Study," 140). The hyphen was also adopted by the journal owing to Bailey's preference for it and "to emphasize the compound term's 'unity,'" according to Kohlstedt, *Teaching Children Science*, 186. This sort of hyphenation, creating new compounds to unify terms, is a stylistic characteristic found throughout Bailey's writing.

26. Maurice Alpheus Bigelow (1872–1955), professor of education at Teachers College, Columbia University, founding editor of *The Nature-Study Review* and secretary-treasurer for the American Nature-Study Society (ANSS), and a leading social hygienist and author of books on sexual education. Bigelow was instrumental in the effort to institutionalize the nature-study movement in the early twentieth century and build academic credibility through the *Review* and ANSS (Kohlstedt, *Teaching Children Science*, 185–192, 202). During his tenure as editor, the *Review* published critics as well as supporters of nature-study, as Bigelow himself seems to have become more ambivalent about the movement. As he shifted away from nature-study, he became more involved in social hygiene, and eventually in the since-debunked and racist field of eugenics. The shift may have been accelerated perhaps by his disapproval of the direction taken by his editorial successor at the *Review*, Fred L. Charles. Bigelow eventually became president of the American Eugenics Society (Kohlstedt, *Teaching Children Science*, 194, 198, 297n48).

Teachers College, where John Dewey came after leaving the University of Chicago (see *Nature-Study Idea*, notes 19–20), led the way in developing urban nature-study curricula for New York City, drawing on the examples of Chicago and Cornell, and in 1906 Bigelow even suggested to Bailey a cooperative program linking their schools in which Teachers College students would attend summer classes at Cornell (Kohlstedt, *Teaching Children Science*, 60–68, 259n13).

The pages of *The Nature-Study Review* "were full of definitions and the boundary work of establishing both connections to and distinctions from agricultural education, social and sexual hygiene, applied science, literature, geography, and other subjects" (Kohlstedt, *Teaching Children Science*, 7). Bailey was among the founding members of the advisory board, and when Bigelow proposed the term "elementary science" be used on the journal's masthead, Bailey objected, and the term was dropped (Kohlstedt, *Teaching Children Science*, 297n53). The journal went through several editors, including Anna Botsford Comstock in 1917, who claimed to have brought subscribers up from twelve hundred to twenty-five hundred in her six years of editorship. In 1923, the editor of *Nature Magazine* offered to merge the two magazines to lift the burden of time and money from Comstock's shoulders, and she agreed on the terms that *The Nature-Study Review* would continue as "a unit in each number of the *Nature Magazine*," that it would remain "the official organ of the American Nature Study Society," and that it would continue to be edited by a member of the society, but Comstock ultimately felt that "*Nature Magazine* did not carry its part of the agreement" and the result of the merger was that "the *Nature Study Review* [sic] was dead and buried" (Comstock, *Comstocks of Cornell*, 428–431; on Comstock, see *Nature-Study Idea*, note 128). *Nature Magazine* itself did not last long, and was succeeded by *Nature and Science Education Review*, which became the official organ of the ANSS in 1928 until it too failed and was succeeded by *Science Education* (Kohlstedt, *Teaching Children Science*, 210). The final official journal of ANSS was simply titled *Nature Study* (Brandwein Institute, "Brief History").

27. The American Nature-Study Society (ANSS) held its first meeting in 1908, in Chicago. It held the meeting in conjunction with the meeting of the American Association for the Advancement of Science in order to bolster nature-study's standing among scientists (Kohlstedt, *Teaching Children Science*, 192). Bailey was elected the society's first president at that meeting, even though he could not be present, evidently in order "to honor Bailey's role as articulate spokesperson and to take advantage of his high visibility" (Kohlstedt 299n83; 193). In its first year, ANSS membership grew to nearly eight hundred, and by 1920, the year that *The Nature-Study Idea* went through its last known print run, membership had grown to twenty-four hundred, with local sections spread across the country but especially clustered in the Northeast and Midwest (Kohlstedt 193–194, 199). The organization persisted and continued to publish a newsletter intermittently through the twentieth and into the twenty-first century, although it struggled with a membership that lowered throughout the later half of its history (Kohlstedt, *Teaching Children Science*, 192–194, 299n83, 199–200). Today, ANSS lays claim to the title of "America's oldest environmental education organization," but its operational status is unclear. The organization's leadership has partnered with the Brandwein Institute to establish the American Nature Study Society Archive Project to preserve and make accessible the society's publications and papers (Brandwein Institute, "Brief History").

28. In his analysis of this critical passage, John P. Azelvandre describes "two crucial elements for the cultivation of the interconnected individual: first, maximum quantity of shared contacts with the world, and second, maximum quality of such contacts – quality being suggested by Bailey's concern for 'sympathy.' Sympathy, when understood as 'fellow-feeling,' suggests the positive apprehension of the values expressed in other entities – the points of contact. It represents a consonance of feeling – a sharing of the ideals of others and an incorporation into self that also includes the transformation of those ideals and the utilization of them in the construction of self. *Eudaimonia* [happiness, flourishing] is thus based on quantity and quality of shared contacts with the world, both human and non-human. [. . .] Cultivation of shared contacts was of course Bailey's primary concern in education" ("Bonds of Sympathy," 290–291).

29. Ralph Waldo Emerson (1803–1882), American essayist and founder of transcendental philosophy. Bailey glosses something of the transcendental outlook with this famous

quotation, while his insistence on the materiality of the wagon qualifies transcendentalism from something like the perspective of the pragmatic philosophers of his generation (such as William James and John Dewey) and the groundedness of agrarianism. The quotation comes from Emerson's Civil War essay "American Civilization," in which he argues for the federal emancipation of the enslaved people of the South: "Hitch your wagon to a star. Let us not fag in paltry works which serve our pot and bag alone. Let us not lie and steal. No god will help. We shall find all their teams going the other way,—Charles's Wain, Great Bear, Orion, Leo, Hercules:—every god will leave us. Work rather for those interests which the divinities honor and promote,—justice, love, freedom, knowledge, utility."

30. From Proverbs 4:7. The larger passage reads:

> ⁵ Get wisdom, get understanding: forget it not; neither decline from the words of my mouth.
> ⁶ Forsake her not, and she shall preserve thee: love her, and she shall keep thee.
> ⁷ Wisdom is the principal thing; therefore get wisdom: and with all thy getting get understanding.
> ⁸ Exalt her, and she shall promote thee: she shall bring thee to honour, when thou dost embrace her.
> ⁹ She shall give to thine head an ornament of grace: a crown of glory shall she deliver to thee. (Prov. 4:5–9 KJV)

31. Followed in the first edition by this passage, cut for the third edition: "We must define nature-study in terms of its purpose, not in terms of its methods. It is not doing this or that. It is putting the child into intimate and sympathetic contact with the things of the external world. Whatever the method, the final result of nature-study teaching is the development of a keen personal interest in every natural object and phenomenon." This then leads directly into what here begins after a section break, "There are two or three fundamental misconceptions [. . .]."

32. This sentence first appears in the third edition. Cf. Bailey, *The Holy Earth*, 7, 10: "Plato, in the 'Republic,' reasoned that the works of the creator must be good because the creator is good. This goodness is in the essence of things; and we sadly need to make it a part in our philosophy of life. The earth is the scene of our life, and probably the very source of it. The heaven, so far as human beings know, is the source only of death; in fact, we have peopled it with the dead. We have built our philosophy on the dead. [. . .]

"But we begin to understand that the best dealing with problems on earth is to found it on the facts of earth. This is the contribution of natural science, however abstract, to human welfare. Heaven is to be a real consequence of life on earth; and we do not lessen the hope of heaven by increasing our affection for the earth, but rather do we strengthen it. Men now forget the old images of heaven, that they are mere sojourners and wanderers lingering for deliverance, pilgrims in a strange land. Waiting for this rescue, with posture and formula and phrase, we have overlooked the essential goodness and quickness of the earth and the immanence of God."

33. "It" was commonly used as a genderless pronoun to refer to people, especially children, at the time that Bailey was writing. The shift toward reserving "it" for inanimate objects came primarily in the twentieth century. Inconsistent usage in *The Nature-Study Idea* loosely tracks with when a passage was written: Bailey appears to prefer the genderless "it" in the 1903 edition, but passages written for the 1909 and 1911 editions reveal that he had moved away from using "it" to refer to people. The discrepancy apparently was not considered significant enough to revise out of older passages like this one.

34. Alfred Russel Wallace (1823–1913), Henry Walter Bates (1825–1892), and John Lubbock (1834–1913), all British naturalists, evolutionists, and notable authors. Wallace and

Bates were well-known for their joint expedition through the Amazon rainforests in the mid-nineteenth century.

35. The nature-study program at Cornell published many such leaflets, distinguishing between "Teachers' Leaflets," which contained a combination of background information in natural history and pedagogical suggestions to aid teachers directly, and "Children's Leaflets," designed to be read and used by students directly. Both types typically featured lush illustrations and were highly literary and companionable in character. The State of New York published a collection of them in book form in 1904 as *Cornell Nature-Study Leaflets*, and later leaflets formed the basis of Anna Botsford Comstock's iconic *Handbook of Nature Study* in 1911 (see "It Is Spirit"; on Comstock, see *Nature-Study Idea*, note 128).

36. The sixth of the Ten Commandments, Exod. 20:13.

37. The "extension movement" in universities seeks to "extend" the university outward, beyond its campus and into communities, in order to, as Bailey puts it in the next sentence, "reach the masses." Bailey helped to pioneer this work during his years at Cornell, and the movement became formalized in the public land-grant university system after the Smith-Lever Act of 1914, which began federal funding of cooperative agricultural extension and which is considered partly to have been an outgrowth of the recommendations laid out by the Commission on Country Life that Bailey chaired under the presidential administration of Theodore Roosevelt. See Peters and Morgan, "Country Life Commission." In line with how Bailey characterizes the "extension idea" here, Peters has argued that "Bailey's vision of agricultural extension work was centered on the provision of education aimed at awakening farmers to a new point of view on life" ("Every Farmer," 190), challenging a historiographic trend to characterize the formative period of the extension system as concerned primarily with increasing agricultural efficiency. See Peters, "Every Farmer Should Be Awakened."

38. Bailey would significantly expand his theory of the "poetic interpretation of nature" two years later in *The Outlook to Nature*—see esp. 12–62—and it receives further treatment later on in *The Nature-Study Idea*, in Part II, Chapter VI. He had been theorizing the relationship of poetry and science, however, at least as early as his unpublished travel narrative of 1888, *Onamanni* (Liberty Hyde Bailey Papers, Box 18, Folders 13–15).

39. Followed in the first edition by this original poem, cut for the third edition:

> Child with the gray-blue eyes
> Gazing so longingly—
> Yonder the great world lies—
> All is unknown to thee!

> Child unwedded to care,
> Softly speedeth the hours—
> Thou buildest castles in air
> And strew'st thy path with flowers.

> *Build on in thy dreaming,*
> *Nor thy fancies are vain;*
> *The best of life's seeming*
> *Are its castles in Spain!*

40. Pegasus, the mythological winged horse of Greek mythology, allowed the hero Bellerophon to ride him in order to defeat the Chimera. Zeus later made Pegasus into the constellation of that name. Bailey invokes the winged steed in reference to his earlier application of Emerson's injunction to "hitch your wagon to a star," implying here that a horse adequate to the task will first be necessary in order to unite star and wagon.

41. Bailey followed a similar paradigm in his university teaching, and he was not afraid to defend his position. According to one biographer: "When a colleague from the classical department observed that it was beneath the dignity of a professor to bring such things as weeds and branches into a lecture hall, Bailey said nothing, but the following day he selected a good-sized log from the woodpile and carried it ostentatiously across the campus and up to his classroom in Morrill Hall" (Dorf, *Liberty Hyde Bailey*, 69).

42. On Agassiz, see *Nature-Study Idea*, note 10. Asa Gray (1810–1888), professor of botany at Harvard University, often considered among the most important American botanists of the nineteenth century, and author of numerous works, including *A Manual of the Botany of the Northern United States* (a standard text that would become known simply as Gray's Manual) and *Darwiniana*. Gray was close friends with Charles Darwin and became an influential apologist for Darwin's theory of evolution, arguing for a theistic view of evolution in which God was the causal force of evolutionary change. After completing his undergraduate work, Bailey studied and worked under Gray in his herbarium at Harvard from 1883 to 1884 before accepting his first faculty position at Michigan Agricultural College. As a child, Bailey had acquired a copy of Gray's *Field, Forest, and Garden Botany* and had consulted it to aid his early plant collecting in the dunes and woods surrounding the town of South Haven, MI (Dorf, *Liberty Hyde Bailey*, 32; Rodgers, *Liberty Hyde Bailey*, 11). After Gray's death, Bailey oversaw a fully revised edition of that book as editor. His copy of Gray's Manual was also very worn and is preserved at the Liberty Hyde Bailey Museum and Gardens in South Haven, Michigan.

43. A reference to "The Brook" by Alfred Lord Tennyson (1809–1892), in which the voice of the brook says:

> I chatter, chatter, as I flow
> > To join the brimming river,
> For men may come and men may go,
> > But I go on forever. (*Complete Poetical Works*, 218.47–50)

44. Followed in the first edition by this passage, cut for the third edition: "To relate the nature-study work to living animals and plants is the fundamental idea in Hodge's ideal, as expressed, for example, in his book, 'Nature-Study and Life.' He holds that the appreciation of inanimate things is a later development in the child-life than an appreciation of objects that are living. He would, therefore, not begin with weathering of rock and formation of soil, combustion and the like, although he would 'not wish to insinuate that the study of living things is all of nature-study.' With this I agree for the very young, and I would study a brook or a fence-corner or a garden-bed or a bird or a plant." For more on Hodge, see *Nature-Study Idea*, note 12.

45. Cf. Bailey, *Holy Earth*: "We have almost forgotten to listen; so great and ceaseless is the racket that the little voices pass over our ears and we hear them not. I have asked person after person if he knew the song of the chipping-sparrow, and most of them are unaware that it has any song. We do not hear it in the blare of the city street, in railway travel, or when we are in a thunderous crowd. We hear it in the still places and when our ears are ready to catch the smaller sounds. There is no music like the music of the forest, and the better part of it is faint and far away or high in the tops of trees" (99).

46. Followed in the first edition by this passage, cut for the third edition: "It is often said that the ignorant man may be as happy as the educated man. Relatively, this is true; absolutely, it is not. A ten-foot well is not so deep as a twenty-foot well; and although the ten-foot well may be full to the brim, it holds only half as much water as the other."

47. This passage was much revised for the third edition. Followed in the first edition by this passage, cut for the third edition: "But, as a matter of fact, nature-study will nearly

always be consecutive in subject-matter because the teacher will feel himself most competent in one or two lines and will devote himself chiefly to them; or the consecutiveness may be that of the seasons, following the wild life of the neighborhood. The gist of it all is that the mere exercises in nature-study are only a means to an end: it is the nature-study spirit, not that exercise nor this, that is to correct and to enliven educational ideals. The given exercise may be secondary to other subjects of the school day, but the point of view—the way of thinking—that it inculcates is fundamental and will pervade the school or the home."

48. Followed in the first edition by a paragraph break and then this passage, cut for the third edition: "If one is to be happy, he must be in sympathy with common things. He must live in harmony with his environment. One cannot be happy yonder nor to-morrow: he is happy here and now, or never. Our stock of knowledge of common things should be great."

49. Sentence preceded in the first edition by this passage, cut for the third edition: "One word from the fields is worth two from the city." The quotation, "God made the country," comes from a famous line in the monumental georgic poem *The Task* (1785) by William Cowper (1731–1800):

> God made the country, and man made the town:
> What wonder then, that health and virtue, gifts
> That can alone make sweet the bitter draught
> That life holds out to all, should most abound
> And least be threaten'd in the fields and groves? (1.749–753)

Bailey includes a meditation on another line from Cowper's *The Task*, "Who loves a garden loves a greenhouse too," in *The Garden Lover* (1928), 69–72; also in *Gardener's Companion*, 199–201.

50. Followed in the first edition by this passage, cut for the third edition:

> I would again emphasize the importance of obtaining our fact before we let loose the imagination, for on this point will largely turn the results—the failure or the success of the movement. We must not allow our fancy to run away with us. If we hitch our wagon to a star, we must ride with mind and soul and body all alert. When we ride in such a wagon, we must not forget to put in the tail-board.
>
> Another most important result of the nature-study movement will be its effect, along with manual-training and other forces, in gradually overturning present systems of schoolwork. The system of memorizing from books will eventually have to go. The pupil will first be put into sympathetic contact with objects, not put into books."

For more on Emerson's injunction to "hitch your wagon to a star," see *Nature-Study Idea*, note 29.

51. Followed in the first edition by this passage, cut for the third edition:

> My own work in nature-study centers chiefly about its value as a means of improving country living. It may tend distinctly toward the improvement of the farmer, and thereby of farming. Go into a potato-growing community and ask the farmers where the roots of the potato plants are—whether above or below the tubers—and you will puzzle them nearly every time. And yet, a knowledge of the position of the roots is essential to the best potato-growing, for upon this position depend in part the principles governing the depth of planting, hilling, and, to some extent, of tilling. At a farmers' meeting in an apple-growing section, I asked how many apple flowers are borne in a cluster. Every man guessed, but no man knew. One man said that the limbs of some of his apple trees had died; he asked me why. I asked him the symptoms: but he did not know as they had any symptoms—they had only died. Had he looked at the limbs? Yes, he had seen them from the barnyard!

Now, I do not care whether nature-study teaches where the potato roots are or not. The point is, that nature-study teaches the importance of actually seeing the thing and then of trying to understand it. The person who actually knows a pussy-willow will know how to become acquainted with the potato-bug. He will introduce himself.

In recent years there has been great activity in disseminating information amongst the farmers. The results have been gratifying. Not only have farmers learned more, but there has been a general uplift in the tone of many rural communities. But the discouraging fact is, that the young people do not often come to the farmers' meetings in any numbers. There will be a constantly recurring crop of ignorance and prejudice. Each crop, to be sure, must be above its predecessor, but yet not living up to the full stature of its opportunities. It is therefore necessary to begin with the new generation—to begin our chimney at the bottom, rather than at the top. People crowd into the cities largely because of the intellectual entertainment that they find there. If their own intellectual horizon is enlarged, they may find entertainment in the country.

The teacher, the clergyman, the progressive merchant or farmer here and there, are the persons that are willing to help along the work of uplifting the rural communities. Education is the only salvation for the farmer—not the development of facts merely, but the development of power through the enlargement of capability. The results will come slowly. We must not be impatient. There are centuries of inertia to be overcome. The best and most permanent things are of slow growth.

An additional paragraph, following this passage, was revised and moved to Part I, Chapter VII.

52. Followed in the first edition by this passage, cut for the third edition, which concludes the chapter (everything that follows in the text of the present edition was added in 1909): "If, in conclusion, I were asked for a condensed statement of the nature-study idea, I should choose the following definition of it by Professor Thomas H. Macbride, of the University of Iowa: 'I should say that by nature-study a good teacher means such study of the natural world as leads to sympathy with it. The keynote, in my opinion, for all nature-study is sympathy. Such study in the schools is not botany; it is not zoölogy; although, of course, not contravening either. But by nature-study we mean such a presentation, to young people, of the outside world that our children learn to love all nature's forms and cease to abuse them. The study of natural science leads, to be sure, to these results, but its methods are long and have a different primary object.'"

Thomas Huston Macbride (1848–1934), American naturalist specializing in botany, mycology, and geology and professor of botany who would become the tenth president of the University of Iowa (1914–1916). He was author of scientific works like *The North American Slime-Moulds* (1899) as well as the textbook *Lessons in Elementary Botany for Secondary Schools* (1895) and the memoir *In Cabins and Sod-Houses* (1928). For a brief biographical sketch, see Mccartney, "Macbride."

53. Cf. Bailey's novel *The Seven Stars* (1923), in which the protagonist concludes at the end of the book that "my aim is the artistic expression of life" (165).

54. Included in this volume under Related Writings. In later editions of the leaflet, which long remained in circulation, the following footnote was added to the first page: "Teacher's Leaflet No. 1, December, 1896. For a discussion of the title of this leaflet and what it signifies pedagogically, consult 'The Integument-Man,' in 'The Nature-Study Idea.' (Doubleday, Page & Co.)" See, e.g., *Cornell Nature-Study Leaflets*, 291.

55. Followed in the first edition by this passage, cut for the third edition: "The method of presentation must first be adapted to the person to be instructed, else the instruction will be of little consequence."

56. The following section contrasting "the child" with "the Integument-Man" was greatly reorganized for clarity in the 1909 edition, and the following was cut from the beginning:

"The Integument-Man sees the little things. The child sees the big things. Ask a child to describe a house, or to draw one."

57. This paragraph links the Integument-Man to the debate that was, at the time of this book's writing, beginning to rage in literary circles between elder naturalist John Burroughs and a host of newer nature writers, including Ernest Thompson Seton, Jack London, and William J. Long, whom Burroughs accused of being "sham naturalists" and misleading children by overly anthropomorphizing the animal protagonists of their stories. Several years later, the then-sitting President Theodore Roosevelt weighed in on the side of Burroughs, describing the young animal-story writers as "nature fakers" and thereby giving the controversy the name by which it is remembered today. Bailey would directly address what he described as the "Burroughs-Long controversy" two years later in *The Outlook to Nature* (1905), 267–270, arguing for a middle ground between extreme pedantism on the one hand and irresponsible romanticism on the other, but he notes that, in general, the animal-story approach of emphasizing the unique traits of individual animals as opposed to the generalized characteristics of the species is "the natural way of knowing the out-of-doors," citing *Black Beauty* and *The Call of the Wild* as examples (268). He calls for more of this "intimate unconscious boy kind of knowledge" to be "put into books" (269). For more on the debate, see Lutts, *The Nature Fakers*.

58. A humorous reference to the passage from the Psalms: "The LORD shall preserve thy going out and thy coming in from this time forth, and even for evermore" (Psalm 121:8 KJV).

59. This chapter appears to be an expansion of the introductory essay, "Paragraphs for the Teacher," that Bailey wrote for his *Botany: An Elementary Text for Schools* (1900). In the first edition of *The Nature-Study Idea*, this opening sentence is preceded by the following, cut for the third edition: "Any one who has listened to discussions in the recent meetings of teachers and scientists must have been impressed with the great prominence which is given to nature-study. The nature-study movement is now, perhaps, the most conspicuous new feature in educational ideals in the secondary and primary schools."

60. While Bailey takes issue with herbarium-collecting here, as an enforced pedagogical method, he himself amassed an extensive herbarium throughout his life that would come to occupy much of his work in retirement. It eventually became part of what he called his "Hortorium," a "repository for things of the garden," which included his 125,000-specimen herbarium, buildings, a test garden, a 3000-volume library, and an 80,000-piece seed and nursery catalog collection, all of which he donated to Cornell University in 1935 and which became a division of the university under his directorship, with his daughter Ethel Zoe Bailey as curator. Today the L. H. Bailey Hortorium Herbarium is one of the largest university-affiliated collections of preserved plant material in North America. On herbaria for children, see, also, the disclaimer that Bailey provides in the final paragraph of this chapter, which was first added to the third edition, and his notes on "collecting" elsewhere in the book. I am grateful to Robert Dirig for clarifying for me some points about the history and scope of the Hortorium.

61. On Asa Gray, see *Nature-Study Idea*, note 42.

62. Followed in the first edition by this passage, cut for the third edition: "It should be related to the experiences of the daily life. It should not be taught for the purpose of making the pupil a specialist: that effort should be retained for the few who develop a taste for special knowledge."

63. Among "our best text-books," Bailey may have had in mind his own, which include examples of most of these types of studies; see *Nature-Study Idea*, note 8.

64. While it appears that this essay was written originally for *The Nature-Study Idea*, Bailey had expressed many similar ideas in his more lengthy experiment station bulletin, "Hints

on Rural School Grounds," published by Cornell in 1899, which reveals the great depth of thought he had given to the issue during his nature-study work across the state of New York.

65. Followed in the first edition by this passage, cut for the third edition: "All this great interest in nature is reacting profoundly on the natural sciences in making them more vital and increasing their application to the daily life. With all its progressiveness, science is yet conservative."

66. On Froebel and Pestalozzi, see *Nature-Study Idea*, note 13.

67. This vignette first appeared in Bailey's leaflet "A Plant at School," published as the February 1903 issue of *Junior Naturalist Monthly*, and it also appeared in the collection *Cornell Nature-Study Leaflets* (1904) as Leaflet 52.

68. In the third edition, Bailey cut a third item here: "(*c*) teaching agriculture and horticulture." Bailey, like Anna Botsford Comstock (*Nature-Study Idea*, note 128), persistently refused to conflate nature-study with the vocational training movement that was then also gaining steam, although he recognized the extent to which the two complemented each other. While nature-study should set the stage for a career in farming and may even especially benefit agricultural communities, Bailey maintained, it should just as well set the stage for any fulfilling life, and the child should not be coerced into any specialty or career. For this reason, he argued for "agricultural nature-study" and not "nature-study agriculture" (quoted in Kohlstedt, *Teaching Children Science*, 104), but see Part I, Chapter VII of this volume for a complication of that equation.

69. Bailey, whose first major academic post was as chair of the Department of Horticulture and Landscape Gardening at Michigan Agricultural College, described these landscaping ideals in his short essay "The Picture in the Landscape," published in *Science* in 1893 and often reproduced in his later works, including *Manual of Gardening* (12–16). Much of the material in this section is also adapted from his 1899 experiment station bulletin "Hints on Rural School Grounds," which lays out the sort of "expert advice" he recommends here.

70. Followed in the first edition by this passage, cut for the third edition: "The improvement of the grounds is the first consideration: that is primarily a question of civic pride."

71. Followed in the first edition by this passage, cut for the third edition: "The ground should be 'good,' well prepared, well tilled."

72. Followed in the first edition by a paragraph break and then this passage, which provides a window into the state of the school garden movement at the time that the first edition was published:

> Just now there is much interest in school-gardening in the United States. This interest is the beginning of a new movement which will take the pupil out-of-doors and to nature, and will relate his school life to his real life. The primary effort should be to arouse the public conscience to the importance of caring for the school premises and to the necessity of bringing the child into sympathy with its environment. Then, here and there, the school-garden, for purposes of definite instruction, will be instituted. In the country districts the school-garden will come slowly, because gardens are so common as to lose their interest, and because the rural schools are often small and weak. Higher ideals of agriculture at home, nature-study in the school, consolidation of weak districts—these are the means that will bring the real school-garden to the rural school.

This then leads into the paragraph that begins the section titled "The larger relations" in the third and subsequent editions. It is worth noting, in this cut passage, that the state of rural school consolidation in the United States was drastically less extensive in 1903 than it has been since the latter part of the twentieth century.

73. Followed in the first edition by this passage, cut for the third edition: "The day is coming when agriculture—under other names, perhaps, and not as a professional subject—will be taught in public schools as a 'culture study.' "

74. Followed in the first edition by this passage, cut for the third edition: "The greater the number of parks the better for the children."

75. Followed in the first edition by this passage, which concludes the chapter, cut for the third edition:

> Some of the specific ways in which our outlook has been extended by the growth of horticulture—which is the growing of plants—may be mentioned:
>
> It has opened our eyes to all the multitude of flowers and ornamental plants.
>
> It has increased our national wealth and has opened the way for large commercial industries.
>
> It has elevated the public taste so that parks and well-kept lawns are now a civic necessity.
>
> It has had much to do with the breadth and spirit of the modern movement that we call nature-study.
>
> It has made plants a part of the home, as books and pictures are. Plant collections stand for culture. Not only do they appeal to the individual who has them, but also to a wide circle of persons, since they are living, growing things and cannot well be hidden.
>
> It has awakened an intrinsic interest in natural objects. People have come to love plants. They like the plant itself as well as its flowers. They know that a plant is worth growing merely because it is a plant. They have come to feel that every animal and plant lives its own life. It has its battles to fight. It contends. Thereby is the individual man carried beyond himself.

76. Sentence preceded in the first edition by this passage, cut for the third edition: "The nature-study idea is fundamental to the evolution of popular education. Therefore it may be applied—in fact, must be applied—to all branches of education." The chapter title was also changed, from "The Agricultural Phase of Nature-Study."

77. For more on the extension movement and university extension, also discussed elsewhere in this chapter, see *Nature-Study Idea*, note 37.

78. William Harold Payne (1836–1907), pedagogue and professor of education, author of books including *Chapters on School Supervision* and *Contributions to the Science of Education*, and, at the University of Michigan, the first chair of a university department of education in the United States. Without a college degree, he began teaching at the age of seventeen, and after several principalships in New York and then Michigan he became president of the Ypsilanti Normal School, then superintendent of schools in Adrian, MI, and finally chair of the new department at the University of Michigan in 1879. From 1887–1901, he served as president of the Peabody Normal College in Nashville, TN (now a part of Vanderbilt University), from which position he thought he could help improve the educational system in the South. While he believed in universal and compulsory free education and was against punitive testing, he doubted the value of nature-study and experiential learning. See Dillingham, "University of Nashville," esp. 330–331. I have not located the source of the quotation.

79. This chapter was largely rewritten for the third edition, beginning after this sentence. By the third edition, Bailey had better integrated the chapter into the style and presentation of the rest of the book, but the first edition provides a fascinating glimpse into the practical work being done by the Cornell nature-study program and how it fit into his larger vision for what would become the country life movement and what he called "the spiritualizing

of agriculture." The remainder of this chapter as it first appeared is included in this volume under Major Sections Restored from the First Edition.

80. It seems unlikely that Bailey meant "power" here in the domineering sense so much as in the sense of the empowerment of a historically oppressed class of people. As he would memorably write in *The Holy Earth* (1915), "For years without number—for years that run into the centuries when men have slaughtered each other on many fields, thinking that they were on the fields of honor, when many awful despotisms have ground men into the dust, the despotisms thinking themselves divine—for all these years there have been men on the land wishing to see the light, trying to make mankind hear, hoping but never realizing. They have been the pawns on the great battlefields, men taken out of the peasantries to be hurled against other men they did not know and for no rewards except further enslavement. They may even have been developed to a high degree of manual or technical skill that they might the better support governments to make conquests. They have been on the bottom, upholding the whole superstructure and pressed into the earth by the weight of it. When the final history is written, the lot of the man on the land will be the saddest chapter" (91–92). He argued that, in the nineteenth and twentieth centuries, this history began to shift as "the man on the bottom began really to be recognized politically," and his hope was that such recognition in the future would be born from "the desire to give the husbandman full opportunity and full justice" and explicitly *not* from "the use that a government can make in its own interest of a highly efficient husbandry" (92).

81. For more on the founding of Cornell's nature-study program, see *Nature-Study Idea*, note 6 and Major Sections Restored.

82. The National Grange of the Order of Patrons of Husbandry, a social and political organization founded in 1867 and devoted to agricultural advocacy and community support. At its height in 1875, the Grange claimed nearly a million members. Grange halls were widespread in rural areas in the early 1900s, much like masonic lodges and other social organizations were. The Grange was notable not only for its impact on U.S. agricultural policy—advocating for what would become the "Granger Laws" to regulate the fares of railroad and grain elevator companies and for what became the Rural Free Delivery program of the U.S. Postal Service, for instance—but also for including women from the beginning, forming economic cooperative endeavors to support farmers, supporting women's suffrage, and emphasizing education and community service. Bailey was a regular speaker at grange hall meetings. For a recent history of the Grange, see Bourne, *In Essentials, Unity*.

83. In the first edition, this chapter was followed by an eighth, short chapter to close out Part I, titled "Review." About half of that chapter was rearranged and worked into Part II, Chapter I for the third edition. To read Part I, Chapter VIII, as it appeared in the first edition, see Major Sections Restored, this volume.

84. An early version of this chapter, under the title "The Point of View Towards Nature," appeared as the first essay in Bailey's 1902 book *Nature Portraits: Studies with Pen and Camera of Our Wild Birds, Animals, Fish and Insects* (1–3). For more on that book, see the essay "It Is Spirit," this volume. Several of Bailey's essays in *Nature Portraits* had appeared previously as editorials in the pages of the magazine *Country Life in America*, and they would all appear in Part II of *The Nature-Study Idea* except for the volume's single poem, "Utility," which he would later include in his collection, *Wind and Weather* (1916), 124–125.

85. Followed in the first edition by this passage, cut for the third edition, which concluded the brief chapter:

> This challenging of the point of view is the theme of the text that I am writing.
>
> Nature-study, properly handled, interprets nature. It does not stop dead with the information that is acquired. It endeavors to understand as well as to see.

86. This paragraph, and the conception of "nature" as "the universal environment" that surrounds us in "city or country or on the sea," anticipates Bailey's later formulation of "the everlasting backgrounds" in his book series The Background Books: The Philosophy of the Holy Earth. In the first volume, *The Holy Earth* (1915), he defines this sense of "backgrounds" as "large environments in which we live but which we do not make [. . .] to which we adjust our civilization, and by which we measure ourselves" (97). His adoption of the term attempts rhetorically to encourage his readers to reorient their attention from narrowly human concerns to this "universal environment" that is the living earth around them. Also in *The Holy Earth*, he offers special consideration of three nonexhaustive examples of background spaces: the "forest primeval," the "open fields," and the "ancestral sea" (97–100, 106–110), partially echoing the phrase "in city or country or on the sea" here. Bailey's understanding that "all things are of kin" in this passage of *The Nature-Study Idea* is rooted in his understanding of evolution, the ethical implications of which he describes as "the brotherhood relation" in *The Holy Earth*, noting there that the "living creation is not exclusively man-centred: it is bio-centric" (25).

87. This chapter previously appeared as the text of Part III in *Nature Portraits* (1902), 17–24, and in that form it also included at the end some of the question-and-answer text that would be incorporated into Part III of *The Nature-Study Idea*.

88. Followed in the first edition by this passage, cut for the third edition: "Their view is necessary in all matters of fact and truth, but not when points of view are concerned."

89. Followed in the first edition by this passage, cut for the third edition: "We are taught, also, that we should develop and strengthen the natural powers."

90. Followed in the first edition by this passage, cut for the third edition, which concludes the chapter: "The way to teach is, after all, mostly a matter of experience and expediency. Things were not made either to be analyzed or collected."

91. This chapter previously appeared as the text of Part IV in *Nature Portraits* (1902), 25–31, and in that form it also included at the end some of the question-and-answer text that would be incorporated into Part III of *The Nature-Study Idea*.

92. Followed in the first edition by this common aphorism, cut for the third edition: "Flowers are fleeting."

93. From the Psalms. The passage reads:

> ³ When I consider thy heavens, the work of thy fingers, the moon and the stars, which thou hast ordained;
> ⁴ what is man, that thou art mindful of him? and the son of man, that thou visitest him? (Psalm 8:3–4 KJV)

94. Pierre Louis Moreau de Maupertuis (1698–1759), French mathematician, philosopher, and author, whose work on heredity is sometimes considered to be a precursor to our modern understanding of genetics and evolution. His principle of least action remains one of his best-known contributions. Gray, a close confidant of Darwin's in the years leading up to the publication of *On the Origin of Species* and later an important mentor to Bailey, mounted this argument against Agassiz in his paper "On the Botany of Japan and its Relations to that of North America and of other Parts of the Northern Temperate Zone" in 1859, shortly before *On the Origin of Species* appeared (see Gray, "Flora of Japan," 135). The work cemented his reputation as one of the world's great botanists (Farlow, "Asa Gray," 171–172; "Asa Gray" 326). For more on Agassiz, see *Nature-Study Idea*, note 10; for more on Gray, see *Nature-Study Idea*, note 42.

95. This chapter previously appeared in *Nature Portraits* (1902), 9–16, which, along with the poem "Utility," which followed it (see *Nature-Study Idea*, note 100), constituted the text of Part II in that book.

96. Followed in the first edition by this passage, cut for the third edition: "We must think of these things as we come and go."

97. The source appears to be *The History of the Propagation and Improvement of Vegetables by the concurrence of Art and Nature* (1660) by Robert Sharrock: "To Trees that bear great heads, and are of a fast and binding bark, such as Cherrie trees, some hard Apples, and other kinds of great fruit-bearing, and other plants, it is esteemed necessary by some to put in more grafts than one, least the sap finding not way enough, the tree receive a check and perish by the disappointment of the sap" (69). I have not located the source of the quotation in the preceding paragraph.

98. Followed in the first edition by this passage, cut for the third edition: "I wish that people might learn to see dandelions."

99. Followed in the first edition by this passage, cut for the third edition: "People want to believe in definite, final, set events."

Charles Darwin (1809–1882), English naturalist and travel writer, author of *The Voyage of H.M.S. Beagle* and *The Origin of Species*, best known for his work advancing the theory of evolution by natural selection. Bailey first encountered Darwin's writings when he was a boy, and they had a profound impact on him; he certainly would have considered himself among Darwin's "followers" (Dorf, *Liberty Hyde Bailey*, 22; Rodgers, *Liberty Hyde Bailey*, 11). Bailey often found that Darwin's theories were popularly misunderstood, though, and in the following passage he provides such a case, in which people insert the idea of "universal adaptation" into the theory of natural selection and thereby find a comfortable (though un-Darwinian) substitute for the unscientific "dogma of special creation."

100. When this chapter appeared in *Nature Portraits*, it was followed by this poem:

UTILITY

In deepest wood
A flow'ret stood
 'Neath unknown skies.
Its petals bright
Ne'er gave their light
 To human eyes.

A wand'ring man
'Neath learning's ban
 Espied the flow'r.
"Ah, little swain
Thy life was vain
 Until this hour."

But Nature knew
Of all that grew
 No thing was vain,
The restless tease
Of busy bees
 Had render'd gain.

When the poem appeared in Bailey's 1916 collection *Wind and Weather*, he added a final stanza:
 As you and me,
 So flower and bee

> Hath life to give;
> Nor pride nor pelf
> Each of itself
> Hath right to live.

101. This chapter previously appeared under the same title as an editorial in the April 1902 issue of *Country Life in America*, 214–215, and, that same year, as part of the text of Part I in *Nature Portraits* (1902), 25–31.

102. Theodore Roosevelt (1858–1919), twenty-sixth president of the United States, prominent conservationist, avid hunter, and author of many books, including *Ranch Life and the Hunting-Trail* (1888) and *African Game Trails* (1910), the latter of which Bailey cites in *The Holy Earth* as providing evidence against the idea that the evolutionary "struggle for existence" could in any way justify war (58). Roosevelt gave his first annual message to Congress on December 3, 1901, and portions of it (including this one) were circulated widely, including in the pages of *Science* (Roosevelt, "Message to the Congress"). While Roosevelt is often remembered as the "father" of the national park system in the United States, Yellowstone was founded as a national park well before his presidency, in 1872, and is considered the world's first national park. For more on the ecological and conservation history of the park, see Schullery, *Searching for Yellowstone*.

103. Bailey lived to see the extinction of the passenger pigeon (*Ectopistes migratorius*), when the last known individual, a captive known as Martha, died in 1913, just over a decade after he wrote this passage. While public opinion had indeed risen to check the wanton eradication of these birds, which at one time were the most abundant bird species on the continent (if not the planet) and were known to darken the sky for days at a time during their yearly migrations, public opinion was too late to save them from the overhunting that was largely driven by the "mere fashion" Bailey indicts here. In his ninetieth birthday speech, Bailey reminisced about learning to trap passenger pigeons as a child from the local Potawatomi, some three hundred of whom he remembered to have lived on the Bailey family's farmland, and who had thus trapped pigeons sustainably for centuries (Ninetieth Birthday Speech, 25–26). His experiences with the Potawatomi, he said then, led him to "pick [. . .] up something of their outlooks" (26). For more on the history of the passenger pigeon and its extinction, see Greenberg, *Feathered River*.

104. The bobolink (*Dolichonyx oryzivorus*) is a migratory songbird common to eastern and central North America. The bobolink's name comes from a loose interpretation of its song, rather than being shorthand for "Robert of Lincoln," but Bryant's poem helped to immortalize the bird. The repeated nonsense lines in the poem seek to imitate the bobolink's vocalizations. Today, bobolinks are fairly common in grasslands, but it is considered a "Priority Bird" by the National Audubon Society, and their numbers have declined some 65% since 1966, owing largely to changing land use and the decline of meadows and hay fields. See Cornell Lab of Ornithology, "Bobolink," and Kaufman, "Bobolink."

105. This chapter previously appeared as the text of Part V in *Nature Portraits* (1902), 33–40, concluding the volume.

106. Followed in the first edition by this passage, cut for the third edition: "Fact is not to be worshiped. The life that is devoid of imagination is dead; it is tied to the earth."

107. Followed in the first edition by this passage, cut for the third edition: "The trouble with much of the sentiment is that it gives us a wrong point of view." Also in the first edition, the first sentence of this paragraph ended with the word "sentiment" rather than "figures of speech."

108. Bailey here again intervenes in the literary debate that would later become known as the "nature faker controversy"; see *Nature-Study Idea*, note 57.

109. Followed in the first edition by this passage, cut for the third edition: "Interest in things themselves should be the primary motive; sentiment comes chiefly as a result. But if there is danger of making sentiment too prominent, there may be equal danger in insisting on a perfunctory scientific point of view."

110. This chapter previously appeared in the December 1901 issue of *Country Life in America*, 37–40. It is the only chapter in Part II not to have previously appeared in *Nature Portraits*. The brook that Bailey describes in this chapter as having frequently visited in his childhood is likely the same brook that still runs through part of the old farm property where he grew up in South Haven, MI, which has since been broken up and sold off. A portion of that property, including the 1858 farmhouse, is preserved today as the Liberty Hyde Bailey Museum and Gardens. The brook lies north of the museum's property and has been at least partly channelized, but it may still be found, south of the First Assembly of God's parking lot and dividing two segments of Bailey Avenue. The City of South Haven and its residents might consider whether the brook that Bailey immortalized here and the remaining stand of woods it stretches into, bordering the museum property to the northeast, ought to be saved from further development and made more accessible to the enjoyment of the community in every season.

111. Followed in the first edition by this sentence, cut for the third edition, which concludes the paragraph: "The winter makes the spring worth while."

112. Followed in the first edition by this passage, cut for the third edition: "Verily, the subjects of which the teacher does not know are useful in the teaching; and then, they are so common!"

113. On "object-teaching," see Part I, Chapter II.

114. Cultural epoch theory emerged in the nineteenth century as a means of rethinking curriculum to follow what child psychologists theorized as a "natural development" of children. Proponents such as Johann Frederich Herbart and G. Stanley Hall (see *Nature-Study Idea*, note 12) argued that a child's development repeats, or "recapitulates," what was seen as the linear, progressive evolution of societies through history—a progression that, for these proponents, always carried a clear Western bias. Such thinkers then organized curricula that matched the child's supposed stage of development to content representing the corresponding stage of supposed societal development: for instance, Herbartian educators in Germany crafted a curriculum that began with epic folklore for young children and progressed to studying the Reformation over the span of eight years. This recapitulation theory built on what the German zoologist and natural philosopher Ernst Haeckel in 1866 called the "biogenetic law," that "ontogeny recapitulates phylogeny"—which is to say that the development of the human individual correlates to the development of the species. This became an important theory for eugenicists, and it led Haeckel, Hall, and others to compare young children to members of "savage" races and to conclude that people racialized as non-white were at lower stages of development and therefore incapable of more advanced learning. Eugenicists even used these ideas to equate the mental capacity of non-white adults to that of very young white children.

Racist eugenic theories like the biogenetic law and recapitulation have been easily disproven many times, but the idea that curriculum should follow a predictable progression of child development altered the course of American education. Theorists like John Dewey (*Nature-Study Idea*, notes 19–20) took a different, "child-centered" approach to developmentalism, arguing that elementary educators should follow each child's individual development and learn from the child what content is appropriate to their stage of growth. In Bailey's answer here, he takes a clear stand against the pedagogical application of recapitulation theory to teaching and curriculum development. His ambivalence to the theory itself could also suggest that he had his doubts about its fundamental merit. For overviews of these topics, see Lassonde, "Developmentalists Tradition"; Kleeberg-Niepage, "Recapitulation Theory"; and Schultz and Schubert, "Cultural Epoch Theory."

115. Followed in the first edition by this passage, cut for the third edition: "I have little sympathy with what is known as 'practical' knowledge as a means of training youth—for that spirit which would teach only those things that can be turned into direct use in money-getting; but I would put the child in contact with its own life [. . .]."

116. This inquiry previously appeared in *Nature Portraits* (1902), appended to the end of the essay, "Science for Science's Sake" (which became the second chapter of Part II in *The Nature-Study Idea* the next year), following the inquiry "But do you think that this nature-study will make investigators?" The order of the two inquiries was reversed in the first edition of *The Nature-Study Idea*, and the latter was cut for the third edition and is included in this volume under Major Sections Restored.

117. Followed in the first edition by this passage, cut for the third edition: "Professor E. B. Titchener writes as follows of what he considers to be the three dangers in nature-study: 'The first is that, in striving for sympathy with nature, we run into sentimentality. The second is that, in avoiding fairy tales, we run into something ten times worse—if indeed fairy tales are bad at all; I mean a pseudo-psychology of the lower animals. And the third is that, in trying to be exceedingly simple, we become exceedingly inaccurate.'"

The quotation comes from the pages of *Science*, in the form of a letter responding to the very same article that Bailey writes off in the first paragraph of *The Nature-Study Idea*, which had been assembled by William Beal to critique Bailey's nature-study philosophy by gathering together the opinions of several "eminent scientific men." See the essay "It Is Spirit," this volume; *Nature-Study Idea*, note 1; and Titchener, "Natural History."

Edward Bradford Titchener (1867–1927) was an English psychologist remembered today for developing the theory of structural psychology. He joined the faculty of the Sage College of Philosophy at Cornell University in 1892 and just three years later became the founding director of the university's Department of Psychology, where he spent the remainder of his career as an institutional colleague of Bailey's. For a recent biographical sketch, see Proctor and Evans, "E. B. Titchener."

118. This inquiry previously appeared in *Nature Portraits* (1902), appended to the end of the essay, "The Extrinsic and Intrinsic Views of Nature" (which became the third chapter of Part II in *The Nature-Study Idea* the next year), following the inquiry "Would you begin by first reading to the child about nature?"

119. This inquiry previously appeared in *Nature Portraits* (1902), appended to the end of the essay, "The Extrinsic and Intrinsic Views of Nature" (which became the third chapter of Part II in *The Nature-Study Idea* the next year).

120. This inquiry previously appeared in *Nature Portraits* (1902), appended to the end of the essay, "The Extrinsic and Intrinsic Views of Nature" (which became the third chapter of Part II in *The Nature-Study Idea* the next year), following the inquiry "Shall we teach the child to collect, and thereby to kill?"

121. Followed in the first edition by this passage, cut for the third edition: "My own love of nature was given direction and purpose by a teacher who knew very little about nature; but she knew how to touch a boy's heart." This teacher was likely Julia Field-King, a rural schoolteacher in South Haven, MI, while Bailey was attending school there as a child, and to whom Bailey dedicated the third and subsequent editions of *The Nature-Study Idea*. On Field-King, see the essay "It Is Spirit," this volume, and Dorf, *Liberty Hyde Bailey*, 20–23.

122. Followed in the first edition by this passage, cut for the third edition: "Our books contain them."

123. Followed in the first edition by this passage, cut for the third edition: "Many investigators are so intent on the accuracy of mere details that they overlook the value of enthusiasm and point of view."

124. Davis's article, titled "A Successful School-Garden: Sketch of the Whittier School-Garden in Virginia," appeared in the March 1903 issue of *Country Life in America*, 192–194. Jean E. Davis is likely Jennie Eliza Davis (1837–1935), better known professionally as *Jane* E. Davis but who appears under the name Jean in the 1900 U.S. Census; she was an editor, author, and educator and identified in census records as white. An 1878 graduate of Vassar College, Davis began teaching mathematics and science at Hampton in 1879. In 1900 she ceased teaching and became full-time editor of the institute's journal, *The Southern Workman*, and in 1902 she took on the directorship of Hampton's new Nature-Study Bureau, which issued a series of leaflets similar to those published by the Cornell program (Tarter, "Jennie Eliza Davis").

The Hampton Institute (now Hampton University), together with the Tuskegee Institute in Alabama (now Tuskegee University), were two of the most high-profile examples of Progressive Era educational institutions dedicated to people of color that attempted to work out programs of nature-study for Black and Indigenous students. Both Hampton and Tuskegee sent students to Cornell to study nature-study and agricultural education. John Spencer (who, as the character "Uncle John," engaged in correspondence with thousands of New York school children in Cornell's Junior Naturalist and similar programs—see "It Is Spirit," and Major Sections, note 3) even set up correspondence between pupils in Tuskegee's training school and students in New York's rural schools (Kohlstedt, *Teaching Children Science*, 106–107).

Despite these partnerships, Kohlstedt finds in her analysis of the programs at Hampton and Tuskegee that they tended to veer away from Bailey's philosophical goal to "put the child in sympathy with nature," often instead emphasizing vocational training in agriculture (*Teaching Children Science*, 105–108). At its best, such a shift was intended to set disenfranchised students on a path of economic empowerment and independence; at its worst, it was assimilationist or could even be interpreted as maintaining a racialized peasant class. Armitage gives a more sympathetic reading of the programs, considering the example of George Washington Carver at Tuskegee, who spearheaded the nature-study program there and sat on the editorial board of *The Nature-Study Review* at Bailey's request and who "encouraged creativity and aesthetic appreciation as well as economic innovation" (Armitage, *Nature Study Movement*, 191, 189–191; Hersey, "Hints and Suggestions," 246; on the *Review*, see *Nature-Study Idea*, note 26). Armitage goes on to argue that the "humanist and creative side of nature study tempered and localized the drive for scientific and economic efficiency" in these programs, and therefore that nature-study reframes a historiography that has tended to focus more squarely on the scientific/economic drive in its analyses of agricultural education efforts in Black and Indigenous communities during the Progressive Era (191). In her analysis of the nature-study programs at the Tuskegee and Hampton Institutes in the context of African American environmentalist history, Dianne D. Glave both acknowledges that nature-study "stabilize[d] the African American workforce that supported the white, Southern way of life" and also argues that "African Americans developed their own nature study programs in tandem with a national movement that was characterized by a curriculum developed in white schools." In these Black-led programs, "teachers and children had the opportunity to practice and revel in a preservationist's appreciation of nature," and "African Americans did all this at schools limited by racism" (*Rooted in the Earth*, 107, 114). For more on the implementation and legacy of nature-study in Progressive Era schools of color, see "It Is Spirit," this volume.

125. Followed in the first edition by a colon and then this passage, cut for the third edition: "this is nature-study, for, to a very great degree, the child is the creature of its environments."

126. Sara May "Sal" Bailey (1887–1936), later Sara Bailey Sailor through marriage, and Ethel Zoe Bailey (1889–1983). Ethel Bailey became an accomplished botanist and an important collaborator of her father's, especially as curator of his herbarium, a title she was officially conferred when the Liberty Hyde Bailey Hortorium Herbarium was donated to Cornell University in 1935 and continued to hold until 1957 (on the Hortorium, see *Nature-Study Idea*, note 60). One of her major contributions in that position was the compilation of one of the largest collections of seed and nursery catalogs in the world (Bates, "Ethel Zoe Bailey"; L. H. Bailey Hortorium Herbarium, "Ethel Z. Bailey").

127. In fact, Bailey would later donate a portion of his farm property, Bailiwick, to a Girl Scouts camp founded by his nature-study colleague Anna Botsford Comstock (*Nature-Study Idea*, note 128). His lake-facing fieldstone house and a remnant apple orchard from the original farm are still in use today as part of the Comstock Adventure Center operated by Girl Scouts of NYPENN Pathways. See also Linstrom, "Land, Labor, Literature," 116–126, 135.

128. Anna Botsford Comstock (1854–1930), educator, artist, author of many books including the bestselling *Handbook of Nature Study* and *The Comstocks of Cornell*, and a central leader of the nature-study movement at Cornell University. A masterful wood engraver, she provided illustrations for many of the publications written by her husband, the entomologist and Cornell professor John Henry Comstock, and received a degree in natural history from Cornell in 1885. She became familiar with the work being done at Oswego with object teaching (*Nature-Study Idea*, note 17) and as early as 1891 ran a summer school field laboratory devoted to outdoor learning. In response to the Panic of 1893, New York designated a Committee for the Promotion of Agriculture, which was charged with investigating the "abandoned farm problem" that was developing in response to the depression as many farmers lost their land. Anna Comstock was appointed to the committee, which in 1894 appointed additional extension funding to Cornell to address the problem and explicitly used the term "nature study" as part of the effort's charge (on extension, see *Nature-Study Idea*, note 37). Comstock was part of the effort to shape Cornell's nature-study program from the beginning, and her reputation as an excellent and engaging teacher as well as theorist grew quickly. In 1897, after Bailey had taken over leadership of the program from Isaac P. Roberts, he hired Comstock as the first woman professor at Cornell, although the Board of Regents overturned her appointment as assistant professor and she worked as an instructor until she was finally promoted in 1913. As Cornell's program rose in prominence through its leaflets and other publications, many of which she authored, Comstock became a national leader in the nature-study movement, and she edited *The Nature-Study Review* from 1917–1923. Kohlstedt argues that "[h]er straightforward commentary, which reflected on both the elegant simplicity and yet the diversity of nature, combined with an acute artistic sensibility made her distinctive among the nature study theorists" (*Teaching Children Science*, 83). *The Handbook of Nature Study* remains in print and in demand as a resource for teachers today, and the unexpurgated edition of her autobiography, *The Comstocks of Cornell*, was just published in 2020 under the editorship of Karen Penders St. Clair. See Kohlstedt, *Teaching Children Science*, 78–84; and Comstock, *Comstocks of Cornell* and *Handbook of Nature Study*, xi–xiv.

129. On The National Grange of the Order of Patrons of Husbandry, which was made up of local "granges" such as Bailey references here (and at which Bailey frequently spoke), see *Nature-Study Idea*, note 82.

130. Milton Pratt Jones (1886–1912), American extension instructor and correspondent with rural students across New York State through Cornell's nature-study program. Jones

began working as an assistant in Cornell's extension department (on extension, see *Nature-Study Idea*, note 37) in the last year of his undergraduate work at Cornell, and on graduating he was promoted to instructor, in which role he wrote letters to rural children in the *Cornell Rural School Leaflet* series. He was stricken with illness at the age of twenty-two and tragically died from the disease three years later ("Milton Pratt Jones"). The essay that Bailey adapts here, titled "Advice to Teachers," appeared in the November 1909 issue of the series. In 1907, Cornell launched the *Cornell Rural School Leaflet* to supplant *Junior Naturalist Monthly* in Cornell's nature-study publishing program, in part out of Bailey's frustration when the New York State Department of Agriculture dropped funding for *Junior Naturalist Monthly* because of the particular popularity of the leaflets among urban teachers in New York City (Kohlstedt, *Teaching Children Science*, 94–95).

131. Scanned copies of these bulletins are available through the digital companion exhibition to this book at www.lhbaileyproject.com.

132. Scanned copies of these bulletins are available through the digital companion exhibition to this book at www.lhbaileyproject.com.

133. The word "museum" at this time could still refer to a simple collection or display of artifacts or specimens, as Bailey means it here, and did not necessarily refer to the kind of public institution typically implied by the term today.

134. According to the Consumer Price Index Inflation Calculator of the U.S. Bureau of Labor Statistics, these numbers in 1913 (the earliest year available in the calculator and the year of the second printing of the fourth edition of *The Nature-Study Idea*) would be about equal to the following, as of May 2021: $137.34 for the Babcock milk test; $20.60 for the "tripod lens magnifying glass"; $34.34 for the terrarium; and $54.94 for the aquarium (U.S. Bureau of Labor Statistics, "CPI Inflation Calculator").

135. Presumably Julia Field-King, to whom Bailey dedicated the third and subsequent editions of the book. For more on Field-King and her relationship to Bailey, see "It Is Spirit," this volume.

136. In *The Training of Farmers* (1909), Bailey writes the following about these new special schools for agricultural training:

> These special schools will undoubtedly be of great value, and they ought to lead the way in a new kind of secondary education; but at the same time we must not forget that we have a public-school system that ought to be developed in these very lines, and it would be a pity to cripple this system by diverting attention elsewhere. We ought not to have duplicate systems of education. These special schools, of whatever plan or organization, should supplement the public-school system, providing facilities for such persons as desire to go further than the public school can take them or who desire quickly to acquire a working knowledge of particular parts of farm life. (168)

137. A scanned copy of this bulletin is available through the digital companion exhibition to this book at www.lhbaileyproject.com.

138. A common trope used to refer to the rural one-room schoolhouse. The phrase has carried nostalgic weight for over a century and is still frequently invoked; see Zimmerman, *Small Wonder: The Little Red Schoolhouse in History and Memory*.

For a sense of the kind of country school Bailey was envisioning that would emphasize hands-on nature-study learning, see the illustrations of Cornell's model rural schoolhouse included in the essay, The Common Schools and the Farm-Youth, this volume under Related Writings.

139. Followed in the first edition by this passage, cut for the third edition: "We are children of nature, and we have never appreciated the fact so much as we do now."

Major Sections Restored

1. On the Junior Naturalist Clubs, see Major Sections, note 3.

2. Mary Farrand Rogers Miller (1868–1971), educator, naturist, and author of books including *The Brook Book* (1901) and *Outdoor Work* (1911). After studying entomology at Cornell, she began working in the nature-study program there as lecturer in 1897, where she continued for six years, regularly contributing to the program's leaflet series (see *Nature-Study Idea*, note 35) and delivering lectures to teachers and farmers' institutes throughout the state (Kohlstedt, *Teaching Children Science*, 93; St. Clair, "Inspirational Voices," 165–166). As described in this passage, Bailey appointed Miller to the position of director of the course in Home Nature-Study in 1902, and that same year he also hired her as assistant editor of *Country Life in America*, a position she held until 1909, several years after Bailey resigned from his editorship (St. Clair, "Inspirational Voices," 165–166). She later became involved in education for the deaf, and she moved to California, where she taught as a lecturer in the extension division of the University of California (Binheim and Elvin, *Women of the West*, 67).

3. John Walton "Uncle John" Spencer (1843–1912), American legislator, fruit farmer, educator, and Master of the New York State Grange ("John Walton Spencer"; Kohlstedt, *Teaching Children Science*, 91; on the Grange, see *Nature-Study Idea*, note 82). Bailey and Anna Botsford Comstock (see *Nature-Study Idea*, note 128) met Spencer on their initial horse-and-buggy tour among rural schools of New York in 1896 and believed he would be of benefit to their nature-study program (Palmer, "Nature Study Philosophy," 40). A leader of the Chautauqua Horticultural Society, Spencer became instrumental in securing the initial appropriation for the nature-study program at Cornell, and Bailey initially brought him into the program to do outreach work with farmers. When Spencer asked to work with young people, however, Bailey agreed, and, as Bailey describes in this passage, Spencer helped form the Junior Naturalist Clubs and quickly became known to thousands of rural school children in and beyond New York with whom he corresponded as "Uncle John." As club membership increased, Spencer required assistance in replying to all the children's letters and began using standardized responses, and, by 1906, Spencer reported a total membership of some thirty thousand children (Kohlstedt, *Teaching Children Science*, 91–92). For more on the work of the Junior Naturalist Clubs, see Bailey's discussion of it later in this chapter and in Nature-Study on the Cornell Plan under Related Writings, this volume. Writing to his publishers at Macmillan in 1896, Bailey described Spencer as "one of the most progressive and intelligent farmers whom I have ever met," and confided that, although he didn't appear frequently in published leaflets, "the man who is really behind this [nature-study] movement is John W. Spencer" (quoted in Colman, *Education & Agriculture*, 123). In her autobiography, Comstock writes of Spencer, "We had stood shoulder to shoulder in our battle to introduce Nature Study in the schools. He [was] brave and full of faith and had often supported me when I might have faltered. When I find Nature Study growing in importance in the schools of the United States I find myself saying to Uncle John, 'Your soul goes marching on'" (*Comstocks of Cornell*, 346).

4. Olly Jasper Kern (1861–1945), school administrator, agricultural educator, and author of books including *The Country School and the Country Child* (1902) and *Among Country Schools* (1906). He became principal of a small village school in 1891 and then superintendent of the schools of Winnebago County, IL, in 1899, a position he kept until 1913. During this period, he became a national voice for school gardens, traveling libraries, agricultural children's clubs, and rural school improvement, and in 1913 he joined the staff of the College of Agriculture at the University of California, Berkeley, until his retirement in 1930 (Crocheron, Butterfield, and Griffin, "Olly Jasper Kern," 103–104). *The Country School and the*

Country Child was the first in an annual series of reports on the work he was overseeing in Winnebago County, and in this passage Bailey may be referring to the descriptions and plans outlined there for district school gardens and for the Winnebago County Farmer Boys' Experiment Club, which formed that year in conjunction with the extension service of the College of Agriculture at the University of Illinois in Urbana, although the specific phrase "district school experiment garden" does not seem to appear in the volume (Kern, *Country School*, see esp. 21–47).

5. David Felmley (1857–1930), educational administrator, mathematics professor, and sixth president (1900–1930) of Illinois State Normal University (now Illinois State University). After receiving his bachelor's degree from Blackburn College in 1881, he became superintendent of schools in Carrollton, IL. In 1890 he became a professor of mathematics at Illinois State Normal University, and in 1900 he assumed the role of president. He believed that all students, not just those planning to attend college, had a right to a high school education, at a time when American high school enrollment was drastically expanding, and he argued that normal schools (teachers' colleges) were better equipped to train primary and secondary teachers than what he saw as elitist private colleges and state universities. To keep graduates abreast of evolving curricula, he oversaw the development of specialized programs in manual arts, domestic science, agriculture, commerce, home economics, and industrial arts (Freed, *Educating Illinois*, 174–204). The first three sentences of this quotation appear in Felmley's 1902 essay "Horticulture in the Elementary Schools," 97; perhaps he quoted himself in correspondence with Bailey.

6. John J. McMahan (ca.1866–1936), American legislator, educational administrator, and lawyer. He was a member of the education committee of South Carolina's constitutional convention of 1895 and served as the state's superintendent of public schools from 1898–1902 (South Carolina Department of Education, "John J. McMahan"). In 1899, an outline of his speech "The Country School Problem" was published as a pamphlet, which argued for reinvestment in rural schools and that universal education "should not aim to prepare for college; it should prepare for life" (3). His advocacy was also driven by the 1895 convention's larger program to disenfranchise Black children, however, who at that time were taking fuller advantage of the new public school system than white children were. After his term as superintendent of South Carolina, he became a member of the state's house of representatives from 1905–1910 and 1915–1916, sitting on the education committee at least part of that time, and served as state insurance commissioner from 1921–1928 (Harlan, *Separate and Unequal*, 170–209, esp. 174–175n12; "J. J. M'Mahan"; South Carolina Department of Education, "John J. McMahan"). My thanks to Susan Smith at the Sumter County Library for pointing me to some of these sources.

Reviews of the Third (1909) Edition

1. In quoting this out of context, the reviewer at the *Times* evidently did not catch the sarcasm of the passage or the argument of the entire chapter.

2. Bailey had corrected this typo, and a number of others that slipped by the copyeditors at Doubleday, by the second edition in 1905.

Related Writings

1. Kohlstedt, *Teaching Children Science*, 93, 269–270n87–88.
2. Colman, *Education & Agriculture*, 185.
3. Colman, *Education & Agriculture*, 185.
4. Bailey, "Garden Fence."

5. The essay continues with a section titled "Retrospect and Prospect after five years' work," which reproduces the same text from the Cornell Experiment Station's *Sixth Report of Extension Work*, 1902, that also appears in Part I, Chapter VII of the first edition of *The Nature-Study Idea*. The reader may consult this text in the present volume, under Major Sections Restored. Following that quotation, Bailey concludes the essay for *Cornell Nature-Study Leaflets* with the following paragraph:

> The literature issued by the Bureau of Nature-Study is of two general types: that which is designed to be of more or less permanent value to the teacher and the school; and that which is of temporary use, mostly in the character of supplements and circulars designed to meet present conditions or to rally the teachers or the Junior Naturalists. The literature of the former type is now republished and is to be supplied gratis to teachers in New York State. The first publication of the Bureau of Nature-Study was a series of teachers' leaflets. This series ran to twenty-two numbers. It was discontinued in May, 1901, because it was thought that sufficient material had then been printed to supply teachers with subjects for a year's work. It was never intended to publish these leaflets indefinitely. Unfortunately, however, some persons have supposed that because these teachers' leaflets were discontinued we were lessening our efforts in the nature-study work. The fact is that later years have seen an intensification of the effort and also a strong conviction on the part of all those concerned that the work has permanent educative value. We never believed so fully in the efficiency of this kind of effort as at the present time.

6. This misinformed concept of "Western" cultural superiority is symptomatic of the dehumanizing tradition in scholarship known as Orientalism, in which, among other prejudices, "The West is [figured as] the actor, the Orient a passive reactor" (Said, *Orientalism*, chapter 1.4). See generally Said, *Orientalism*. In spite of these cultural blinders, Bailey would later spend time in China studying regional agricultural practices there and argue that American agriculture, often characterized by short-sighted practices contributing to soil exhaustion, had much to learn from Chinese methods of soil conservation in farming, which had in some cases been employed on the same land for thousands of years. See Bailey, *What Is Democracy?*, esp. 125–175, and his preface to F. H. King, *Farmers of Forty Centuries; or, Permanent Agriculture in China, Korea and Japan*, iii–iv.

7. The April 1907 issue of *The Nature-Study Review* (vol. 3, no. 4, 113–115) ran the following announcement about the construction of the Cornell Rural School House, which similarly featured the building's floorplan, along with two photographs, one of which is the clear model for the first illustration included above in The Common Schools and the Farm-Youth.

THE CORNELL RURAL SCHOOL HOUSE

In a letter referring to this new building described below Professor Bailey explained its purpose as follows: "I have built this schoolhouse primarily for the purpose of raising the whole qusestion of the rural school and its efficiency. Whatever the merits of this particular building may be, the question is up for discussion. One may go from Maine to Minnesota and see practically the same kind of rural school building, and it is the same type of building as was in use fifty years or more ago. In cities and towns the new ideas are expressed in new school buildings, new churches, new residences, and new kinds of stores and shops. I think it is quite useless to talk about the reorganization of the school curriculum without talking, at the same time, about the reorganization of the building in which the work is to be done."

The New York State College of Agriculture at Cornell University has erected a small rural school-house on its grounds, to serve as a suggestion in school-house architecture and to contain a real rural school as a part of its nature-study department.

The prevailing rural school-house is a building in which pupils sit to study books. It ought to have a room in which pupils do personal work with both hands and mind. The essential feature of this new school-house, therefore, is a work-room. This room occupies one-third of the floor space. Perhaps it would be better if it occupied two-thirds of the floor-space. If the building is large enough, however, the two kinds of work could change places in this school-house.

It has been the purpose to make the main part of the building about the size of the average rural school-house, and then to add the work-room as a wing or projection. Such a room could be added to existing school buildings; or, districts in which the building is now too large, one part the room could be partitioned off as a work-room.

It is the purpose, also, to make this building artistic, attractive and home-like to children, sanitary, comfortable, and durable. The cement-plaster exterior is handsomer and warmer than wood, and on expanded metal lath it is durable. The interior of this building is very attractive.

The picture shows the building as just completed, before the grading of the grounds. School-gardens and play-grounds are being made at one side.

The cost has been as follows: Contract price for buildings complete, including heater in cellar, blackboards, and two outhouses with metal drawers, $1800; tinting of walls $25.00; curtains $16.56; furniture and supplies $141.75; total $1,983.31. In rural districts, the construction might be completed at less cost. The average valuation of rural school buildings and sites in New York State in 1905 was $1,833.63.

The building is designed for twenty-five pupils in the main room. The folding doors and windows in the partition enable one teacher to manage both rooms. The openings between school-room and work-room are fitted with glazed swing sash and folding doors, so that the rooms may be used either singly or together, as desired. The work-room has a bay window facing south and fitted with shelves for plants. Slate black-boards of standard school heights fill the spaces about the rooms between doors and windows. The building is heated by hot air; vent flues of adequate sizes are also provided so that the rooms are thoroughly heated and ventilated.

On the front of the building and adding materially to its picturesque appearance, is a roomy veranda with simple square posts, from which entrance is made directly into the combined vestibule and coat-room and from this again by two doors into the school-room.

Inquiries about the construction details of this school building may be addressed to L. H. Bailey, Director College of Agriculture, Ithaca, N. Y.

8. This passage seems to evoke the myth of the "vanishing Indian," a myth that functioned to hide the truth that Indigenous people in the Americas were not innocently "vanishing" but were victims of systematic efforts at forced removal, assimilation, and genocide. For a classic study, see Dippie, *Vanishing American*. Bailey's own writing elsewhere, especially in *Onamanni*, indicates that he should have known better than to fall back on this old romantic trope in this essay. The Potawatomi settlements around South Haven had indeed "vanished" from the landscape since his childhood, but it was because of continuing settler colonial campaigns following the violent removal earlier in the nineteenth century that became known as the "Potawatomi Trail of Death" (see Willard and Campbell, *Potawatomi Trail of Death*). This passage can also be read against the grain of the myth of vanishment, to the extent that Bailey also made clear (in *The Nature-Study Idea*; see II.V) that he knew that the passenger pigeon's absence from the woods was directly due to the violence of European settlement—indeed, none of the changes glossed in this sentence can be attributed to nonhuman, "natural" causes.

9. Much wartime propaganda during the first World War pitted British/American "culture" against German "*Kultur*," and many observers and intellectual commentators understood the war primarily as a clash of cultures. See, e.g., Temkin, "Culture vs. *Kultur*." Rather than elevate culture over *Kultur* here, Bailey critiques the ideology of cultural superiority generally in his suggestion to move past even the English form of the word. While many American progressives supported American intervention in the war, Bailey maintained an early and sustained opposition to the war generally, as can be seen in *The Holy Earth* (1915), *Universal Service* (1918), and *What Is Democracy?* (1918).

10. Yet another reference to the animal stories that were the subject of the nature faker controversy; see *Nature-Study Idea*, note 57. "Lobo the Wolf" is the protagonist of Ernest Thompson Seton's short story "Lobo, the King of Currumpaw," from *Wild Animals I Have Known*, 1898; "Black Beauty the Horse" is the protagonist of Anna Sewell's novel *Black Beauty: His Grooms and Companions, the Autobiography of a Horse*, 1877; and "The Cat that Walked by His Wild Lone" is the protagonist of Rudyard Kipling's short story "The Cat that Walked by Himself" (described elsewhere in the story as walking "by his wild lone"), from *Just So Stories*, 1902.

11. Thomas Henry Huxley (1825–1895), English biologist and anthropologist, often referred to as "Darwin's bulldog" for his adamant support for Darwin's theories of evolution, and author of the famous *Evidence as to Man's Place in Nature* (1863).

WORKS CITED

[Agassiz, Louis.] "Methods of Study in Natural History." *Atlantic Monthly* 9, no. 51 (Jan. 1862): 1–13. HathiTrust, https://hdl.handle.net/2027/chi.34938353.

Allen, Kathleen. "The Women of Cornell and the Nature Study Movement, 1880–1930." PhD diss., Union Institute and University, 2021.

American Nature Study Society and Brandwein Institute. "The American Nature Study Society Archive Project." American Nature Study Society Archive. Accessed September 12, 2022. https://brandwein.org/anss/.

Ardoin, Nicole M., Alison W. Bowers, and Estelle Gaillard. "Environmental Education Outcomes for Conservation: A Systematic Review." *Biological Conservation* 241 (January 2020): 1–13. ScienceDirect, https://doi.org/10.1016/j.biocon.2019.108224.

Armitage, Kevin C. *The Nature Study Movement: The Forgotten Popularizer of America's Conservation Ethic*. Lawrence: University Press of Kansas, 2009.

"Asa Gray." *Proceedings of the American Academy of Arts and Sciences* 23, part 2 (1888): 321–343.

Azelvandre, John P. "Forging the Bonds of Sympathy: Spirituality, Individualism and Empiricism in the Ecological Thought of Liberty Hyde Bailey and Its Implications for Environmental Education." PhD diss., New York University, 2001. ProQuest, https://search-proquest-com.proxy.library.nyu.edu/docview/304710819.

Bailey, Ethel Z. Interview by Gould P. Colman, June 27–October 22, 1963. Liberty Hyde Bailey Museum and Gardens, South Haven, MI.

Bailey, L[iberty]. H[yde, Jr]. *Beginners' Botany*. 1908. New York: Macmillan, 1909.

——. *Beginners' Botany*. 1908. Authorized in the provinces of Nova Scotia and New Brunswick. Macmillan's Canadian School Series. Toronto: Macmillan, 1916.

——. *Beginners' Botany*. 1915. Western edition. Adapted for Canadian schools by B. J. Hales. Authorized for use in the provinces of Manitoba and Saskatchewan. Macmillan's Canadian School Series. Toronto: Macmillan, 1925.

——. *Beginners' Botany*. Authorized by the Minister of Education for Ontario. Toronto: Macmillan, 1921. Internet Archive, ark:/13960/t5h99kb2t.

——. *Botany: An Elementary Text for Schools*. 1900. New York: Macmillan, 1911.

——. *Botany for Secondary Schools: A Guide to the Knowledge of the Vegetation of the Neighborhood*. 1900. New York: Macmillan, 1916.

——. *The Country-Life Movement in the United States*. The Rural Outlook Set 4. New York: Macmillan, 1911.

——, ed. *Cyclopedia of American Agriculture: A Popular Survey of Agricultural Conditions, Practices and Ideals in the United States and Canada*. 4 vols. New York: Macmillan, 1907–1909.

——, ed. *Cyclopedia of American Horticulture: Comprising Suggestions for Cultivation of Horticultural Plants, Descriptions of the Species of Fruits, Vegetables, Flowers and Ornamental Plants Sold in the United States and Canada, together with Geographical and Biographical Sketches*. 4 vols. New York: Macmillan, 1900–1902.

——. "The Garden Fence." 1885. In *The Liberty Hyde Bailey Gardener's Companion*, edited by John A. Stempien and John Linstrom, 227–261. Ithaca, NY: Cornell University Press, 2019.

——. *The Garden Lover*. The Background Books: The Philosophy of the Holy Earth 7. New York: Macmillan, 1928.

——. *Ground-Levels in Democracy*. Ithaca, NY, published by the author,1916.

——. "Hints on Rural School Grounds." Cornell University Experiment Station, Bulletin 160 (January 1899). HathiTrust, https://hdl.handle.net/2027/uiug.30112019742219.

——. *The Holy Earth*. 1915. Centennial ed., edited by John Linstrom. The Background Books: The Philosophy of the Holy Earth 1. Berkeley: Counterpoint Press, 2015.

——. *How a Squash Plant Gets Out of the Seed*. Teacher's Leaflets for Use in the Rural Schools 1 (December 1, 1896). In *Second Report upon Extension Work in Horticulture*, by L. H. Bailey. Cornell University Agricultural Experiment Station, Bulletin 122 (December 1896): 496–500. HathiTrust, https://hdl.handle.net/2027/uiug.30112019741815.

——. Interviews with George H. M. Lawrence, October 21, 1951–October 10, 1952. Corrected transcription by Frank Dennis, Jane Taylor, and Daniel Weinstock, 2007. Electronic file, Microsoft Word. Liberty Hyde Bailey Museum and Gardens, South Haven, MI.

——. *Lessons with Plants: Suggestions for Seeing and Interpreting Some of the Common Forms of Vegetation*. 1897. 2nd ed. New York: Macmillan, 1899.

——. *The Liberty Hyde Bailey Gardener's Companion: Essential Writings*. Edited by John A. Stempien and John Linstrom. Ithaca, NY: Cornell University Press, 2019.

——. *Manual of Gardening: A Practical Guide to the Making of Home Grounds and the Growing of Flowers, Fruits, and Vegetables for Home Use*. 1910. Twelfth printing, rev. New York: Macmillan, 1925.

——. *Nature Portraits: Studies with Pen and Camera of Our Wild Birds, Animals, Fish and Insects.* New York: Doubleday, Page, 1902.

——. "The Nature-Study Idea." *Country Life in America* 1, no. 4 (February 1902): 128–129.

——. *The Nature-Study Idea: An Interpretation of the New School-Movement to Put the Young into Relation and Sympathy with Nature.* 1903. 3rd ed., rev. New York: Macmillan, 1909.

——. *The Nature-Study Idea: An Interpretation of the New School-Movement to Put the Young into Relation and Sympathy with Nature.* 1903. 4th ed., rev. The Rural Outlook Set 2. New York: Macmillan, 1911. Later printing, The Rural Outlook Set 2. New York: Macmillan, 1920.

——. "The Nature-Study Idea: Being an Account of How the Term Originated and What It Really Means." *Our Day* 22, no. 5 (May 1903): 3–4. Google Books, https://books.google.com/books?id=1GTDx8QgRIUC&pg=RA4-PA3&lpg=RA4-PA3&dq=#v=onepage&q&f=false.

——. *The Nature-Study Idea: Being an Interpretation of the New School-Movement to Put the Child in Sympathy with Nature.* New York: Doubleday, Page, 1903. Internet Archive, ark:/13960/t0ht2h17h.

——. *The Nature-Study Idea: Being an Interpretation of the New School-Movement to Put the Child in Sympathy with Nature.* 1903. [2nd ed.] New York: Doubleday, Page, 1905.

——. "The Nature-Study Movement." In *Cornell Nature-Study Leaflets: Being a Selection, with Revision, from the Teachers' Leaflets, Home Nature-Study Lessons, Junior Naturalist Monthlies and Other Publications from the College of Agriculture, Cornell University, Ithaca, N.Y., 1896–1904,* 21–29. Nature-Study Bulletin no. 1. Albany: State of New York–Department of Agriculture, 1904.

——. "The New Hunting." *Country Life in America* 1, no. 6 (April 1902): 214–215.

——. [Ninetieth Birthday Speech.] In *Words Said about a Birthday: Addresses in Recognition of the Ninetieth Anniversary of the Natal Day of Liberty Hyde Bailey,* pamphlet: 24–36. Delivered at Cornell University, April 29, 1948.

——. *Onamanni: A Gardener's Vacation.* 1886–1899. Liberty Hyde Bailey Papers, #21-2-3342, Box 18, Division of Rare and Manuscript Collections, Cornell University Library, Ithaca, NY.

——. "An Outlook on Winter." *Country Life in America* 1, no. 2 (December 1901): 37–40.

——. *The Outlook to Nature.* New York: Macmillan, 1905.

——. *The Outlook to Nature.* 1905. New and rev. ed. New York: Macmillan, 1911.

——. "The Picture in the Landscape." *Science* 22, no. 563 (November 17, 1893): 267–268. JSTOR, https://www.jstor.org/stable/1768004.

——. Preface to *Farmers of Forty Centuries; or, Permanent Agriculture in China, Korea and Japan,* by F. H. King. 1911. Emmaus, PA: Rodale, n.d.

——, ed. *The Principles of Agriculture: A Text-Book for Schools and Rural Societies.* 1898. 5th ed. New York: Macmillan, 1902.

——. *The Seven Stars.* The Background Books: The Philosophy of the Holy Earth 5. New York: Macmillan, 1923.

———. *The State and the Farmer.* The Rural Outlook Set 3. New York: Macmillan, 1908.

———. *Talks Afield about Plants and the Science of Plants.* Boston: Houghton, Mifflin, 1885.

———. *The Training of Farmers.* New York: Century, 1909.

———. *Universal Service.* 1918. The Background Books: The Philosophy of the Holy Earth 3. Ithaca, NY, published by author, 1919.

———. *What Is Democracy?* 1918. The Background Books: The Philosophy of the Holy Earth 4. New York: Macmillan, 1923.

———. *What Is Nature-Study?* 2nd ed. Teacher's Leaflets for Use in the Public Schools 6 (June 1, 1897): 49–52. Tenth Annual Report of the Agricultural Experiment Station: Ithaca, N.Y. 1897. New York: Wynkoop Hallenbeck Crawford, 1898. HathiTrust, https://hdl.handle.net/2027/hvd.hxhqxx.

———. *Wind and Weather.* 1916. The Background Books: The Philosophy of the Holy Earth 2. Ithaca, NY, published by author, 1919.

Bailey, L. H., and Walter M. Coleman. *First Course in Biology.* 1908. New York: Macmillan, 1909.

Bates, David M. "Ethel Zoe Bailey, 1889–1983." *Baileya* 23, no. 1 (January 1989): 1–4. HathiTrust, https://hdl.handle.net/2027/uc1.l0071958730.

Beal, W[illiam]. J. "What Is Nature Study?" *Science* 15, no. 390 (June 20, 1902): 991–992. JSTOR, https://www.jstor.org/stable/1629221.

———. "What Is Nature Study?" *Science* 16, no. 414 (December 5, 1902): 910–913. JSTOR, https://www.jstor.org/stable/1628587.

Binheim, Max, and Charles A. Elvin, eds. *Women of the West: A Series of Biographical Sketches of Living Eminent Women in the Eleven Western States of the United States of America.* Los Angeles: Publishers Press, 1928. HeinOnline, https://heinonline-org.proxy.library.nyu.edu/HOL/Page?handle=hein.peggy/wwbiske0001&collection=peggy.

Bourne, Jenny. *In Essentials, Unity: An Economic History of the Grange Movement.* Athens, OH: Ohio University Press, 2017. ProQuest Ebook Central, https://ebookcentral-proquest-com.proxy.library.nyu.edu/lib/nyulibrary-ebooks/detail.action?docID=4816245.

Boyden, Arthur Clarke. *Albert Gardner Boyden and the Bridgewater State Normal School: A Memorial Volume.* Histories of Bridgewater State University 3. Bridgewater, MA: Arthur H. Willis, 1919. Bridgewater State University Virtual Commons, https://vc.bridgew.edu/bsu_histories/3.

———. *The History of Bridgewater Normal School.* Histories of Bridgewater State University 2. Bridgewater, MA: Bridgewater Normal Alumni Association, 1933. Bridgewater State University Virtual Commons, https://vc.bridgew.edu/bsu_histories/2.

———. *Nature Study by Months: Part I. For Elementary Grades.* 3rd ed. Boston: New England Publishing, 1898. HathiTrust, https://hdl.handle.net/2027/hvd.32044097028369.

Brandwein Institute. "Brief History of the American Nature Study Society." American Nature Study Society Archive. Accessed January 17, 2022. https://brandwein.org/anss/anss-history/.

Brooks, William Keith. "Biographical Memoir of Alpheus Hyatt. 1838–1902." *Biographical Memoirs* 6:310–324. Washington, DC: National Academy of Sciences, 1909. HathiTrust, https://hdl.handle.net/2027/uc1.b000974295.

Bryant, William Cullen. *The Poetical Works of William Cullen Bryant*, vol. 2, edited by Parke Godwin. Vol. 4 of *The Life and Works of William Cullen Bryant*. New York: D. Appleton, 1883. HathiTrust, https://hdl.handle.net/2027/njp.32101075688208.

——. "Robert of Lincoln." *Putnam's Monthly; A Magazine of American Literature, Science, and Art* 5, no. 30 (June 1855): 576–577. HathiTrust, https://hdl.handle.net/2027/umn.31951002805979d.

Colman, Gould P. *Education & Agriculture: A History of the New York State College of Agriculture at Cornell University*. Ithaca, NY: Cornell University, 1963.

Comstock, Anna Botsford. *The Comstocks of Cornell—The Definitive Autobiography*. Edited by Karen Penders St. Clair. Ithaca, NY: Cornell University Press, 2020.

——. *Handbook of Nature Study*. 1911. Ithaca, NY: Cornell University Press, 1986.

Cornell Lab of Ornithology. "Bobolink." In *All about Birds*. Last modified 2019. https://www.allaboutbirds.org/guide/Bobolink/overview.

Cornell Nature-Study Leaflets: Being a Selection, with Revision, from the Teachers' Leaflets, Home Nature-Study Lessons, Junior Naturalist Monthlies and Other Publications from the College of Agriculture, Cornell University, Ithaca, N.Y., 1896–1904. Nature-Study Bulletin no. 1. Albany: State of New York–Department of Agriculture, 1904.

Cowper, William. *The Task, A Poem, In Six Books*. 1785. Eighteenth-Century Poetry Archive, edited by Alexander Huber. https://www.eighteenthcenturypoetry.org/works/o3795-w0010.shtml.

Crocheron, B. H., H. M. Butterfield, and F. L. Griffin. "Olly Jasper Kern, Agricultural Education: Berkeley." In *University of California: In Memoriam, 1943–1945*. University of California (System) Academic Senate, 1943–1945. Online Archive of California, http://www.oac.cdlib.org/view?docId=hb696nb2rz&brand=oac4&doc.view=entire_text.

Cruikshank, Kathleen Anne. "The Rise and Fall of American Herbartianism: Dynamics of an Educational Reform Movement." PhD diss., University of Wisconsin–Madison, 1993. ProQuest, https://www.proquest.com/pqdtglobal/docview/304086719/fulltextPDF/D1FFE19695FC42FBPQ/1?accountid=12768.

Darwin, Charles. *Journal of Researches into the Natural History and Geology of the Countries Visited During the Voyage of H. M. S. Beagle Round the World, under the Command of Capt. Fitz Roy, R.N.* New York: Harper and Brothers, 1848. HathiTrust, https://hdl.handle.net/2027/hvd.32044011908688.

——. *On the Origin of Species by Means of Natural Selection, or the Preservation of Favoured Races in the Struggle for Life*. London: John Murray, 1859. HathiTrust, https://hdl.handle.net/2027/mdp.39015063447794.

Davis, Jean E. "A Successful School-Garden: Sketch of the Whittier School-Garden in Virginia." *Country Life in America* 3, no. 5 (March 1903): 192–194.

Dewey, John. *Democracy and Education*. 1916. New York: Free Press, 1966.

——. *The Public and Its Problems*. 1927. Chicago: Swallow Press, 1954.

——. *The School and Society.* 1899. Rev. ed., Chicago: University of Chicago Press, 1916.

Dillingham, George A., Jr. "The University of Nashville, a Northern Educator, and a New Mission in the Post-Reconstruction South." *Tennessee Historical Quarterly* 37, no. 3 (Fall 1978): 329–338. JSTOR, https://www.jstor.org/stable/42625882.

Dippie, Brian W. *The Vanishing American: White Attitudes and U.S. Indian Policy.* Middletown, CT: Wesleyan University Press, 1982.

Dorf, Philip. *Liberty Hyde Bailey: An Informal Biography.* Ithaca, NY: Cornell University Press, 1956.

Doris, Ellen. "The Practice of Nature-Study: What Reformers Imagined and What Teachers Did." EdD diss., Harvard University, 2002.

"Dr. Piez, 81, Dead in Oswego, Manual Training Pioneer." *Post-Standard*, Syracuse, NY, June 9, 1946. Newspaper Archive, https://newspaperarchive.com/syracuse-post-standard-jun-09-1946-p-27.

Emerson, Ralph Waldo. "American Civilization." *Atlantic*, April 1862. https://www.theatlantic.com/magazine/archive/1862/04/american-civilization/306548/.

"Ethel Z. Bailey Horticultural Catalogue Collection." L. H. Bailey Hortorium Herbarium. Last modified March 2010. http://bhort.bh.cornell.edu/catalogs.htm.

Farlow, W. G. "Memoir of Asa Gray: 1810–1888." *Biographical Memoirs* 3 (1895): 161–175. Washington, DC: National Academy of Sciences.

Felmley, David. "Horticulture in the Elementary Schools." *Elementary School Teacher* 3, no. 2 (October 1902): 96–102. HathiTrust, https://hdl.handle.net/2027/uc1.b3096352.

Fine, Eve. "Medical Education." In *Encyclopedia of Chicago*, edited by Janice L. Reiff et al. Chicago History Museum, Newberry Library, Northwestern University, 2005. http://www.encyclopedia.chicagohistory.org/pages/805.html.

"Frank O. Payne." *School* 33, no. 26 (February 23, 1922), New York: 422. HathiTrust, https://hdl.handle.net/2027/umn.31951000765639o.

Freed, John B. *Educating Illinois: Illinois State University, 1857–2007.* Virginia Beach: Illinois State University and Donning, 2009. Illinois State University, Milner Library, ISU ReD: Research and eData, https://ir.library.illinoisstate.edu/eil/1.

Fuldner, Carl. "Evolving Photography: Naturalism, Art, and Experience, 1889–1909." PhD diss., University of Chicago, 2018. ProQuest, https://www.proquest.com/docview/2161312778/38B147E83502435EPQ/1?accountid=12768.

Glave, Dianne D. *Rooted in the Earth: Reclaiming the African American Environmental Heritage.* Chicago: Lawrence Hill Books, 2010.

Gray, Asa. *Darwiniana: Essays and Reviews Pertaining to Darwinism.* New York: D. Appleton, 1876. HathiTrust, https://hdl.handle.net/2027/uc2.ark:/13960/t3vt1h946.

——. *Field, Forest, and Garden Botany: A Simple Introduction to the Common Plants of the United States East of the 100th Meridian, Both Wild and Cultivated.* Revised and extended by L. H. Bailey. New York: American Book Company, 1895.

——. "The Flora of Japan." [Extract from "Memoir on the Botany of Japan, and Its Relations to That of North America, and of Other Parts of the Northern Temperate Zone."] In *Scientific Papers of Asa Gray.* Selected by Charles Sprague Sargent, vol. 2, 125–141. Boston: Houghton, Mifflin, 1889. HathiTrust, https://hdl.handle.net/2027/hvd.rsm8n1.

——. *A Manual of the Botany of the Northern United States, from New England to Wisconsin and South to Ohio and Pennsylvania Inclusive, (The Mosses and Liverworts by Wm. S. Sullivant,) Arranged According to the Natural System; with an Introduction, Containing a Reduction of the Genera to the Linnaean Artificial Classes and Orders, Outlines of the Elements of Botany, a Glossary, Etc.* Boston: James Munroe and Company, 1848. HathiTrust, https://hdl.handle.net/2027/hvd.hw2o6q.

Greenberg, Joel. *A Feathered River across the Sky: The Passenger Pigeon's Flight to Extinction.* New York: Bloomsbury, 2014.

Harlan, Louis R. *Separate and Unequal: Public School Campaigns and Racism in the Southern Seaboard States, 1901–1915.* Chapel Hill: University of North Carolina Press, 1958. HathiTrust, https://hdl.handle.net/2027/mdp.39015003478057.

Hawken, Paul. *Blessed Unrest: How the Largest Movement in the World Came into Being and Why No One Saw It Coming.* London: Viking, 2007.

Hersey, Mark. "Hints and Suggestions to Farmers: George Washington Carver and Rural Conservation in the South." *Environmental History* 11, no. 2 (April 2006): 239–268. JSTOR, https://www.jstor.org/stable/3986231.

History of the First Half Century of the Oswego State Normal and Training School, Oswego, New York. Oswego: Radcliffe, 1913. HathiTrust, https://hdl.handle.net/2027/msu.31293106769213.

Hodge, Clifton F. *Nature Study and Life.* Boston: Ginn, 1902. HathiTrust, https://hdl.handle.net/2027/uc1.$b17256.

Huxley, Thomas H. *Evidence as to Man's Place in Nature.* New York: D. Appleton, 1863. HathiTrust, https://hdl.handle.net/2027/hvd.32044011695350.

Intergovernmental Panel on Climate Change (IPCC). *Climate Change and Land: An IPCC Special Report on Climate Change, Desertification, Land Degradation, Sustainable Land Management, Food Security, and Greenhouse Gas Fluxes in Terrestrial Ecosystems.* Geneva: IPCC, 2019. Accessed August 19, 2022. https://www.ipcc.ch/srccl/.

Jack, Zachary Michael. "Introducing Sower and Seer, Liberty Hyde Bailey." In *Liberty Hyde Bailey: Essential Agrarian and Environmental Writings.* Edited by Zachary Michael Jack. Ithaca, NY: Cornell University Press, 2008.

Jackman, Wilbur S. *Nature Study for the Common Schools.* New York: Henry Holt, 1891. HathiTrust, https://hdl.handle.net/2027/mdp.39015030970597.

——. *Nature Study for Grammar Grades: A Manual for Teachers and Pupils below the High School in the Study of Nature.* 1898. New York: Macmillan, 1899. HathiTrust: https://hdl.handle.net/2027/coo1.ark:/13960/t4xh0507w.

Jane Taylor Collection. UA.17.292, Michigan State University Archives and Historical Collections, East Lansing, MI.

"J. J. M'Mahan: Former State Superintendent of Education Dies Today." *Sumter Daily Item*, Sumter, SC, January 4, 1936. Newspapers.com, https://theitem.newspapers.com/image/668821318.

"John Walton Spencer Dies at City Hospital." *Cornell Daily Sun* 33, no. 29, October 25, 1912. Cornell Daily Sun Keith R. Johnson '56 Archive, Cornell University Library, https://cdsun.library.cornell.edu/?a=d&d=CDS19121025.1.1&e=-------en-20--1--txt-txIN-------.

Jones, Milton Pratt. *Advice to Teachers.* Cornell Rural School Leaflet 3, no. 3 (November 1909): 34–40. HathiTrust, https://hdl.handle.net/2027/nyp.33433008211645.

Kammen, Carol. *Part & Apart: The Black Experience at Cornell, 1865–1945.* Ithaca, NY: Cornell University Library, 2009.

Kaufman, Kenn. "Bobolink." *Guide to North American Birds.* National Audubon Society. Accessed June 23, 2021. https://www.audubon.org/field-guide/bird/bobolink.

Kellogg, Amos M. *Elementary Psychology.* New York: E. L. Kellogg, 1894. HathiTrust, https://hdl.handle.net/2027/loc.ark:/13960/t5x64bg5z.

——. *Pestalozzi: His Educational Work and Principles.* 1891. New York: E. L. Kellogg, 1894. HathiTrust, https://hdl.handle.net/2027/uc1.a0003707650.

——. *School Management: A Practical Guide for the Teacher in the School Room.* The New Education [series]. New York: E. L. Kellogg, 1880. HathiTrust, https://hdl.handle.net/2027/mdp.39015062739308.

Kern, O. J. *Among Country Schools.* Boston: Ginn, 1906. HathiTrust, https://hdl.handle.net/2027/mdp.39015062314250.

——. *The Country School and the Country Child: Winnebago County, Illinois.* Rockford, IL, 1902. HathiTrust, https://hdl.handle.net/2027/uc1.b2908921.

Kimmerer, Robin Wall. "Weaving Traditional Ecological Knowledge into Biological Education: A Call to Action." *BioScience* 52, no. 5 (May 2002): 432–438. Oxford Academic, https://doi.org/10.1641/0006-3568(2002)052[0432:WTEKIB]2.0.CO;2.

Kipling, Rudyard. *Just So Stories for Little Children.* London: Macmillan, 1902. HathiTrust, https://hdl.handle.net/2027/uc1.31822006039622.

Kleeberg-Niepage, Andrea. "Recapitulation Theory." In *The SAGE Encyclopedia of Children and Childhood Studies,* edited by Daniel Thomas Cook, vol. 1, 1351–1353. SAGE Publications, 2020. https://www.doi.org/10.4135/9781529714388.n499.

Kohlstedt, Sally Gregory. *Teaching Children Science: Hands-On Nature Study in North America, 1890–1930.* Chicago: University of Chicago Press, 2010. ProQuest, https://ebookcentral-proquest-com.proxy.library.nyu.edu/lib/nyulibrary-ebooks/detail.action?docID=534586.

Kuo, Ming, and Catherine Jordan. Editorial: "The Natural World as a Resource for Learning and Development: From Schoolyards to Wilderness." *Frontiers in Psychology* 10 (July 31, 2019): 1763. https://doi.org/10.3389/fpsyg.2019.01763.

Kuo, Ming, Michael Barnes, and Catherine Jordan. "Do Experiences with Nature Promote Learning? Converging Evidence of a Cause-and-Effect Relationship." *Frontiers in Psychology* 10 (February 19, 2019): 305. https://doi.org/10.3389/fpsyg.2019.00305.

Lassonde, Cynthia A. "Developmentalists Tradition." In *Encyclopedia of Curriculum Studies,* edited by Craig Kridel, 286. SAGE Publications, 2010. https://www.doi.org/10.4135/9781412958806.n159.

Liberty Hyde Bailey Papers. #21-2-3342, Division of Rare and Manuscript Collections, Cornell University Library.

Linstrom, John. "Land, Labor, Literature: Ecospheric Critique from the Margins of the Progressive Era." PhD diss., New York University, 2021. ProQuest, https://www.proquest.com/pqdtglobal/docview/2564835985/25E70D5D4F414B0APQ/1?accountid=12768.

Livingstone, David N. *Nathaniel Southgate Shaler and the Culture of American Science.* Tuscaloosa: University of Alabama Press, 1987. HathiTrust, https://hdl.handle.net/2027/mdp.39015012941087.

Lord, Russell. *The Care of the Earth*. New York: Mentor, 1963.

Louv, Richard. *The Last Child in the Woods*. Chapel Hill: Algonquin, 2005.

——. *The Nature Principle*. Chapel Hill: Algonquin, 2011.

——. *Our Wild Calling*. Chapel Hill: Algonquin, 2019.

——. "Outdoors for All." *Sierra*, May–June 2019.

Lurie, Edward. *Louis Agassiz: A Life in Science*. Chicago: University of Chicago Press, 1960.

Lutts, Ralph H. *The Nature Fakers: Wildlife, Science & Sentiment*. Golden, CO: Fulcrum, 1990.

Macbride, Thomas H[uston]. *In Cabins and Sod-Houses*. Iowa City: State Historical Society of Iowa, 1928. HathiTrust, https://hdl.handle.net/2027/mdp.39015027934226.

——. *Lessons in Elementary Botany for Secondary Schools*. 1895. Boston: Allyn and Bacon, 1896. HathiTrust, https://hdl.handle.net/2027/hvd.32044107234627.

——. *The North American Slime-Moulds: Being a List of All Species of Myxomycetes Hitherto Described from North America, Including Central America*. New York, Macmillan, 1899. HathiTrust, https://hdl.handle.net/2027/mdp.39015069534405.

Macmillan Company records. Manuscripts and Archives Division, New York Public Library.

Martin, George H. *The Evolution of the Massachusetts Public School System: A Historical Sketch*. International Education Series 29. New York: D. Appleton, 1894. HathiTrust, https://hdl.handle.net/2027/coo1.ark:/13960/t96691r8r.

——. *A Text Book on Civil Government in the United States*. New York: A. S. Barnes, 1875. HathiTrust, https://hdl.handle.net/2027/hvd.32044097048532.

Mary Baker Eddy Papers. Mary Baker Eddy Library, Boston, MA.

Mccartney, David. "Macbride, Thomas Huston." *The Biographical Dictionary of Iowa*, edited by David Hudson, Marvin Bergman, and Loren Horton. Iowa City: University of Iowa Press, University of Iowa Libraries. Accessed September 14, 2022. http://uipress.lib.uiowa.edu/bdi/DetailsPage.aspx?id=240.

McMahan, John J. "The Country School Problem: Outline of an Address Delivered before the State Teachers' Association at Harris Lithia Springs, S. C., July 17, 1899, by John J. McMahan, State Superintendent of Education." Pamphlet. Accession 1270—M622 (675). Louise Pettus Archives and Special Collections, Winthrop University.

Menand, Louis. *The Metaphysical Club*. New York: Farrar, Straus and Giroux, 2001.

Miller, Mary Rogers. *The Brook Book: A First Acquaintance with the Brook and Its Inhabitants through the Changing Year*. 1901. New York: Doubleday, Page, 1902. HathiTrust, https://hdl.handle.net/2027/chi.086751760.

——. *Outdoor Work*. The Children's Library of Work and Play. Garden City, NY: Doubleday, Page, 1911.

"Milton Pratt Jones." *Cornell Countryman* 9, no. 9 (June 1912): 310. HathiTrust, https://hdl.handle.net/2027/chi.105099584.

Minton, Tyree G. "The History of the Nature-Study Movement and Its Role in the Development of Environmental Education." EdD diss., University of Massachusetts, 1980. Doctoral Dissertations 1896–February 2014, https://scholarworks.umass.edu/dissertations_1/3600.

"Mr. Amos M. Kellogg." *Journal of Education* 76, no. 14 (October 10, 1912): 379–380. HathiTrust, https://hdl.handle.net/2027/hvd.32044102790110.

Morgan, Paul A., and Scott J. Peters. "The Foundations of Planetary Agrarianism: Thomas Berry and Liberty Hyde Bailey." *Journal of Agricultural and Environmental Ethics* 19, no. 5 (2006): 443–468.

"Nature Study or Nature-Study." *Nature-Study Review: Devoted to All Phases of Nature-Study in Elementary Schools* 1, no. 3 (May 1905): 140.

No Child Left Inside Act of 2008. H.R. 3036/S. 1775. 110th Cong. (2007–2008). Congress.gov, https://www.congress.gov/bill/110th-congress/house-bill/3036.

No Child Left Inside Act of 2022. H.R. 7486/S. 4041. 117th Cong. (2021–2022). Congress.gov, https://www.congress.gov/bill/117th-congress/senate-bill/4041/text? r=1&s=1.

"Old Educators Never Die . . . They Just Keep On Keeping On." *Michigan Education Journal* 28 (May 1951): 479–482, 517. Jane Taylor Collection, UA.17.292, Michigan State University Archives and Historical Collections, East Lansing, MI.

Palmer, E. Laurence. *The Cornell Nature Study Philosophy.* Cornell Rural School Leaflet 38, no. 1 (September 1944).

Payne, Frank Owen. *Geographical Nature Studies for Primary Work in Home Geography.* New York: American Book, 1898. HathiTrust, https://hdl.handle.net/2027/uc2. ark:/13960/t3222sm65.

——. *Manual of Experimental Botany.* New York: American Book, 1912. HathiTrust, https://hdl.handle.net/2027/hvd.32044097027247.

——. *One Hundred Lessons in Nature Study around My School.* New York: E. L. Kellogg, 1895. HathiTrust, https://hdl.handle.net/2027/uc2.ark:/13960/t3610x30b.

Payne, William H. *Chapters on School Supervision: A Practical Treatise on Superintendence; Grading; Arranging Courses of Study; The Preparation and Use of Blanks, Records, and Reports; Examinations for Promotion, Etc.* Cincinnati: Wilson, Hinkle, 1875. HathiTrust, https://hdl.handle.net/2027/uiug.30112112043549.

——. *Contributions to the Science of Education.* 1886. New York: Harper and Brothers, 1887. HathiTrust, https://hdl.handle.net/2027/loc.ark:/13960/t1pg2h960.

Peters, Scott J. " 'Every Farmer Should Be Awakened': Liberty Hyde Bailey's Vision of Agricultural Extension Work." *Agricultural History* 80, no. 2 (spring 2006): 190–219. JSTOR, http://www.jstor.org/stable/3744806.

Peters, Scott J., and Paul A. Morgan. "The Country Life Commission: Reconsidering a Milestone in American Agricultural History." *Agricultural History* 78, no. 3 (summer 2004): 289–316. JSTOR, http://www.jstor.org/stable/3744708.

Pretty, Jules. *Agri-Culture: Reconnecting People, Land, and Nature.* London: Earthscan, 2002.

Proctor, Robert W., and Rand Evans. "E. B. Titchener, Women Psychologists, and the Experimentalists." *American Journal of Psychology* 127, no. 4: 501–526. JSTOR, https://doi.org/10.5406/amerjpsyc.127.4.0501.

Rankin, Louise Spieker. Unpublished partial biography of Liberty Hyde Bailey. Louise Spieker Rankin papers, #1438, Division of Rare and Manuscript Collections, Cornell University Library. Typescript.

Rockefeller, Steven. *John Dewey: Religious Faith and Democratic Humanism.* New York: Columbia University Press, 1991.

Rodgers, Andrew Denny, III. *Liberty Hyde Bailey: A Story of American Plant Sciences.* 1949. New York: Hafner, 1965.

Roosevelt, Theodore. *African Game Trails: An Account of the African Wanderings of an American Hunter-Naturalist*. 1910. New York: Cooper Square, 2001.

——. "Extracts from President Roosevelt's Message to the Congress." *Science* 14, no. 363: 907–912. JSTOR, https://www.jstor.org/stable/1627679.

——. *Ranch Life and the Hunting-Trail*. 1888. New York: Century, 1911. HathiTrust, https://hdl.handle.net/2027/hvd.32044087511929.

Ross, Dorothy. *G. Stanley Hall*. Chicago: University of Chicago Press, 1972.

Said, Edward W. *Orientalism*. 1978. New York: Vintage, 1994. EBSCOhost, http://proxy.library.nyu.edu/login?url=https://search.ebscohost.com/login.aspx?direct=true&db=nlebk&AN=842875&site=ehost-live&ebv=EK&ppid=Page-__-1.

Schullery, Paul. *Searching for Yellowstone: Ecology and Wonder in the Last Wilderness*. 1997. Boston: Houghton Mifflin, 1999.

Schultz, Brian D., and William H. Schubert. "Cultural Epoch Theory." In *Encyclopedia of Curriculum Studies*, edited by Craig Kridel, 165. SAGE Publications, 2010. https://www.doi.org/10.4135/9781412958806.n97.

Schulze, Robin G. *The Degenerate Muse: American Nature, Modernist Poetry, and the Problem of Cultural Hygiene*. Oxford: Oxford University Press, 2013.

Seton, Ernest Thompson. *Wild Animals I Have Known: Being the Personal Histories of Lobo, Silverspot, Raggylug, Bingo, the Springfield Fox, the Pacing Mustang, Wully and Redruff*. 1898. New York: Schocken Books, 1966.

Sewell, Anna. *Black Beauty: His Grooms and Companions, the Autobiography of a Horse*. London: Jarrold and Sons, [1877]. HathiTrust, https://hdl.handle.net/2027/uc2.ark:/13960/t4qj79244.

Sharrock, Robert. *The History of the Propagation and Improvement of Vegetables by the concurrence of Art and Nature: Shewing the several ways for the Propagation of Plants usually cultivated in England, as they are increased by Seed, Off-sets, Suckers, Truncheons, Cuttings, Slips, Laying, Circumposition, the several ways of Graftings and Inoculations; as likewise the methods for Improvement and best Culture of Field, Orchard, and Garden Plants, the means used for remedy of Annoyances incident to them; with the effect of Nature, and her manner of working upon the several Endeavors and Operations of the Artist*. Oxford: Tho. Robinson, 1660. Biodiversity Heritage Library, https://doi.org/10.5962/bhl.title.25528.

Sheldon, Edward Austin. *Autobiography of Edward Austin Sheldon*. Edited by Mary Sheldon Barnes. New York: Ives-Butler, 1911. HathiTrust, https://hdl.handle.net/2027/mdp.39015011831990.

Sheldon, E. A., arr. *Lessons on Objects, Graduated Series; Designed for Children between the Ages of Six and Fourteen Years, Containing, also, Information on Common Objects*. New York: Ivison, Blakeman, 1863. HathiTrust, https://hdl.handle.net/2027/msu.31293010955700.

South Carolina Department of Education. "John J. McMahan." Last modified 2021. https://ed.sc.gov/newsroom/former-state-superintendents-of-education/john-j-mcmahan/.

"South Haven is Home Old Friends Are Told," *South Haven Tribune* [?], June 18, 1930. Clipping found in *Liberty Hyde Bailey*, compiler unknown, handbound scrapbook of newspaper clippings. Collection of the Liberty Hyde Bailey Museum and Gardens, South Haven, MI.

St. Clair, Karen Penders. "Finding Anna: The Archival Search for Anna Botsford Comstock." PhD diss., Cornell University, 2017. ECommons, Cornell University, https://doi.org/10.7298/X44X55ZB.

——. "Inspirational Voices in Early Botanical Education." *Plant Science Bulletin* 65, no. 3 (Fall 2019): 161–171. Plant Science Bulletin Archive, Botanical Society of America, https://cms.botany.org/file.php?file=SiteAssets/publications/psb/issues/PSB-2019-65-3.pdf.

Szarkowski, John. *Liberty Hyde Bailey and the Survival of the Unlike.* Unpublished manuscript, private collection of the Estate of John Szarkowski, care of Nina Szarkowski Jones.

Tarter, Brent. "Jennie Eliza Davis (1857–1935)." In *Dictionary of Virginia Biography.* Library of Virginia, 2016. Accessed June 30, 2021. https://www.lva.virginia.gov/public/dvb/bio.php?b=Davis_Jennie_Eliza.

Temkin, Moshik. "Culture vs. *Kultur,* or a Clash of Civilizations: Public Intellectuals in the United States and the Great War, 1917–1918." *Historical Journal* 58, no. 1 (2015): 157–182. Cambridge Core, https://doi.org/10.1017/S0018246X14000594.

Tennyson, Alfred Lord. *The Complete Poetical Works of Tennyson.* Edited by W. J. Rolfe. Cambridge, MA: Houghton Mifflin, 1898. HathiTrust, https://hdl.handle.net/2027/pst.000029837960.

Titchener, E. B. "Natural History in England." *Science* 16, no. 417 (December 26, 1902): 1032–1033. HathiTrust, https://hdl.handle.net/2027/uc1.b000254715.

U.S. Bureau of Labor Statistics. "CPI Inflation Calculator." U.S. Department of Labor. Accessed July 2, 2021. https://www.bls.gov/data/inflation_calculator.htm.

Washington, Booker T. "The Negro Farmer." In *Farm and Community.* 1909. Vol. 4 of *Cyclopedia of American Agriculture: A Popular Survey of Agricultural Conditions, Practices and Ideals in the United States and Canada,* edited by L. H. Bailey, 2nd ed., 106–108. New York: Macmillan, 1910.

——. *Working with the Hands: Being a Sequel to 'Up from Slavery' Covering the Author's Experiences in Industrial Training at Tuskegee.* 1904. Lexington, KY: Library of Congress, 2017.

Westbrook, Robert B. *John Dewey and American Democracy.* Ithaca, NY: Cornell University Press, 1991.

White, Monica. *Freedom Farmers: Agricultural Resistance and the Black Freedom Movement.* Chapel Hill: University of North Carolina Press, 2018. ProQuest Ebook Central, https://ebookcentral-proquest-com.proxy.library.nyu.edu/lib/nyulibrary-ebooks/detail.action?docID=5574865.

Willard, Shirley, and Susan Campbell. *Potawatomi Trail of Death—1838 Removal from Indiana to Kansas.* Rochester, IN: Fulton County Historical Society, 2003.

Williams, Dilafruz R. "Garden-Based Education." In *Oxford Research Encyclopedia of Education,* edited by George Noblit. New York: Oxford University Press, 2018. Oxford Research Encyclopedias, https://doi.org/10.1093/acrefore/9780190264093.013.188.

Williams, Dilafruz R., and Jonathan D. Brown. *Learning Gardens and Sustainability Education: Bringing Life to Schools and Schools to Life.* New York: Routledge, 2012.

Williams, Dilafruz R., and P. Scott Dixon. "Impact of Garden-Based Learning on Academic Outcomes in Schools: Synthesis of Research between 1990 and 2010." *Review of Educational Research* 83, no. 2 (June 1, 2013): 211–235. SAGE Journals, https://doi.org/10.3102/0034654313475824.

Wilson, Edward O. *Biophilia: The Human Bond with Other Species*. Cambridge, MA: Harvard University Press, 1984.

Wu, Tim. *The Attention Merchants*. New York: Knopf, 2016.

Zimmerman, Jonathan. *Small Wonder: The Little Red Schoolhouse in History and Memory*. New Haven, CT: Yale University Press, 2009. JSTOR, https://www-jstor-org.proxy.library.nyu.edu/stable/j.ctt1npmc4.

INDEX

Page numbers in *italics* refer to illustrations.

Central School, South Haven, Michigan, 20–24, *21*, 25, 56
The Century Magazine, 232
Charles, Fred L., 311n26
Chicago, 7, 26
Chicago Institute, 309nn19–20
child-centered learning, 23–24, 325n114
child development, 164–65, 325n114
children. *See* students
Children and Nature Network, xiii, 12
children's organizations, 16–17, 55.
    See also Junior Naturalist clubs
"Child's Realm" (poem), vi
China, 332n6
citizenship and civic engagement, 16, 59, 81, 117–18. *See also* democracy
Clark University, 84, 254, 306n12
classification, 81, 106, 133, 268, 271
climate change, xiii–xiv, 2–3, 11, 12, 17
Coleman, Walter M., 305n8
colleges and universities: agricultural, 121, 126, 187, 193, 265, 329n136; land-grant, xii, xv, 1, 47, 232, 314n37; nature-study in, 35, 79–80; preparation for, 110, 271; students at, 132, 216. *See also* academic disciplines; extension programs; *specific colleges and universities*
The College Speculum (newspaper), 36
Colman, Gould P., 233
Comenius, John Amos, 84, 307n13, 311n24
Commission on Country Life, xii, 53, 232–33, 314n37
common schools (public schools): college preparation, 271; defined, 304n2; discipline in, 25, 182, 264; farm youth and, 265–78; nature-study in, 5–7, 10–17, 24, 26, 29, 38, 40, 79, 119, 233; official nature-study programs, 267. *See also* elementary schools; high schools; rural schools; students; teachers; urban schools
"The Common Schools and the Farm-Youth" (essay), 10, 16, 17, 232–33, 265–78, 332n7

community organization, 7. *See also* rural life and communities
complexity, 2, 9, 130, 190, 195, 213–14, 258
composition, 183–84, 259, 263
Comstock, Anna Botsford: biography, 328n128; as editor of *The Nature-Study Review,* 234, 312n26; as first woman professor at Cornell, 38, 328n128; *Handbook of Nature-Study,* 51, 55, 232, 314n35, 328n128; leadership in nature-study movement, xi, xiii, 7, 19, 20, 24, 28, 184–85, 231, 305n6, 308n18, 309n20, 328n127, 328n128, 330n3; photographs of, *ii, 39*; as poetry editor at *Country Life in America,* 31; vocational training and, 319n68
Comstock, John Henry, *39*, 328n128
Comstock Adventure Center, 56, 328n127
The Comstocks of Cornell (ed. Karen Penders St. Clair), 328n128
conclusions (drawn from observation), 1, 23, 26, 90, 98, 122, 141, 182, 235n, 241, 243, 247, 252, 255, 286, 290
connectedness, xiv, 10–11, 15, 108, 162, 322n86
conservation, 7, 52
conservation education, 12
Cook County Normal School (Chicago), 85, 308n19
cooperation, xiv–xv, 179, 206, 208, 232, 311n26
cooperative agriculture extension, 314n37. *See also* extension programs
Cornell Countryman, 226
Cornell Nature-Study Leaflets (1904), 54–55, 231–32, 314n35, 319n67. *See also* leaflets
Cornell Rural School Leaflet series, 329n130
Cornell University: Bailey Hall, 233; Civic Ecology Lab, 54; Department of Home Economics, 38; L. H. Bailey Hortorium Herbarium, 318n60,

summer camp, 182
summer vacation, 181–82
superficiality, 5, 131, 133, 170–71
sustainability, xv, 1–2, 6, 12, 46–47, 324n103
syllabi, 187, 267–73
sympathy, 37, 44, 46–47, 59, 88–89, 100, 117, 120–22, 164, 172–73, 176, 180, 186, 200–201, 221–22, 230, 245, 254, 256, 258–59, 261–62, 269, 312n28, 316n48, 317n52. *See also* love of nature; spirit

Taft, William Howard, 53
Talks Afield about Plants and the Science of Plants (1885), 29
Taylor, Jane L., 301n11
teachers: effectiveness of, 93, 103–5, 163, 166, 174–75, 195, 244–45, 302n36, 309n19; emergence of nature-study and, 6–7, 24, 28–29, 42, 77; enthusiasm of, 104, 131, 166–67, 175, 248, 257, 327n123; female, 7, 20, 38, 39, 56, 306n10; happiness of, 4; male, 100, 204, 306n10; as overburdened, 171–72; power of, 59, 130, 214; in rural schools, 123–24, 262; social progress and, 26; starting nature-study lessons, 166–67; traveling professionals, 175. *See also* pedagogical methods
Teachers College, Columbia University, 26, 87, 311n26
teachers' institutes, 28, 206, 262
"The Teacher's Interpretation of Nature" (chapter), 24, 129–30
teacher training, 54, 93, 100, 174–75, 193; Agassiz's methods, 306n10; at Cornell Nature-Study Summer School, 39; home nature-study course, 202–3; inquiries and answers about nature-study, 161–95; laboratories and, 169; pedagogical theories and, 81–82; rural schoolteachers, 200–204
technical education, 176–77

Tennyson, Alfred Lord, 315n43
textbooks: by Bailey, 29, 55, 305n8; criticisms of, 5, 170, 316n50; in elementary education, 90; recitation from, 20–21; for rural schools, 125, 212; selection of, 173–74; subject matter in, 168; used in nature-study, 173–74
Thoreau, Henry David, xv, 33
thoroughness, 169–70, 257
Titchener, Edward Bradford, 326n117
Tocqueville, Alexis de, xiv
tools, 100, 190
The Training of Farmers (1909), 329n136
transcendentalism, 33, 312n29
The Tribune Farmer (newspaper), 19
Tuskegee Institute, 48–49, 327n124

understanding, 12, 14–16, 28, 59, 88, 98, 130, 133, 170, 172, 192, 214, 230, 252–53, 269, 313n30
uniformity, 292. *See also* standardized education
Universal Service (1918), 234
universities. *See* colleges and universities
University of Chicago, 251; School of Education, 309nn19–20, 311n26
urban life, 26–27, 36, 264, 299n12
urban schools, 7, 232, 260
uses, 33, 48, 139–42, 220
"Utility" (poem), 33, 321n84, 322n95, 323n100

Van Rensselaer, Martha, 38
Verrill, E. A., 253
vocational training, 319n68, 327n124

Wallace, Alfred Russel, 90, 313n34
Washington, Booker T., 7, 49
watershed education, 12
wealth inequality, xiii, 2
weather, 97, 101, 121, 157, 190–91, 280
Western culture, 269, 332n6, 334n9
"What Is Agricultural Education?" (essay), 231

www.ingramcontent.com/pod-product-compliance
Lightning Source LLC
Chambersburg PA
CBHW030729280326
41926CB00086B/584